MW00564456

CHEMISTRY OF PYROTECHNICS

CHEMISTRY OF PYROTECHNICS

Basic Principles and Theory

JOHN A. CONKLING

Department of Chemistry
Washington College
Chestertown, Maryland

Best regards,
John A. Conkling
November, 1985

MARCEL DEKKER, INC. New York and Basel

Library of Congress Cataloging in Publication Data

Conkling, John A., [date]
Chemistry of pyrotechnics.

Includes bibliographies and index.
1. Fireworks. I. Title.
TP300.C66 1985 662'.1 85-7017
ISBN 0-8247-7443-4

Warning: Formulas in this book relate to mixtures, some or all of which may be highly volatile and could react violently if ignited by heat, spark, or friction. High-energy mixtures should *never* be prepared or handled by anyone untrained in proper safety precautions. All work in connection with pyrotechnics and explosives should be done only by experienced personnel and only with appropriate environmental safeguards. The publisher and the author disclaim all responsibility for injury or damage resulting from use of any formula or mixture described in this book; each user assumes all liability resulting from such usage.

MARCEL DEKKER, INC.
270 Madison Avenue, New York, New York 10016

Current printing (last digit):
10 9 8 7 6 5 4 3 2 1

PRINTED IN THE UNITED STATES OF AMERICA

PREFACE

Everyone has observed chemical reactions involving pyrotechnic mixtures. Beautiful 4th of July fireworks, highway distress signals, solid fuel boosters for the Space Shuttle, and the black powder used by muzzle-loading rifle enthusiasts all have a common technical background.

The chemical principles underlying these high-energy materials have been somewhat neglected in the twentieth century by academic and industrial researchers. Most of the recent work has been goal-oriented rather than fundamental in nature (e.g., produce a deeper green flame). Many of the significant results are found in military reports, and chemical fundamentals must be gleaned from many pages of test results.

Much of today's knowledge is carried in the heads of experienced personnel. Many of these workers acquired their initial training during World War II, and they are presently fast approaching (if not already past) retirement age. This is most unfortunate for future researchers. Newcomers have a difficult time acquiring the skills and knowledge needed to begin productive experiments. A background in chemistry is helpful, but much of today's modern chemistry curriculum will never be used by someone working in pyrotechnics and explosives. Further, the critical education in how to safely mix, handle, and store high-energy materials is not covered at all in today's schools and must be acquired in "on-the-job" training.

This book is an attempt to provide an introduction to the basic principles of high-energy chemistry to newcomers and to serve as a review for experienced personnel. It can by no means substitute for the essential "hands on" experience and training necessary to safely work in the field, but I hope that it will be a helpful companion. An attempt has been made to keep chemical theory simple and directly applicable to pyrotechnics and explosives. The level approaches that of an introductory college course, and study of this text may prepare persons to attend professional meetings and seminars dealing with high-energy materials and enable them to intelligently follow the material being presented. In particular, the International Pyrotechnic Seminars, hosted biannually by the Illinois Institute of Technology Research Institute in conjunction with the International Pyrotechnics Society, have played a major role in bringing researchers together to discuss current work. The Proceedings of the nine seminars held to date contain a wealth of information that can be read and contemplated by persons with adequate introduction to the field of high-energy chemistry.

I would like to express my appreciation to Mr. Richard Seltzer of the American Chemical Society and to Dr. Maurits Dekker of Marcel Dekker, Inc. for their encouragement and their willingness to recognize pyrotechnics as a legitimate branch of modern chemistry. I am grateful to Washington College for a sabbatical leave in 1983 that enabled me to finalize the manuscript. I would also like to express my thanks to many colleagues in the field of pyrotechnics who have provided me with data as well as encouragement, and to my 1983 and 1984 Summer Chemistry Seminar groups at Washington College for their review of draft versions of this book. I also appreciate the support and encouragement given to me by my wife and children as I concentrated on this effort.

Finally, I must acknowledge the many years of friendship and collaboration that I enjoyed with Dr. Joseph H. McLain, former Chemistry Department Chairman and subsequently President of Washington College. It was his enthusiasm and encouragement that dragged me away from the norbornyl cation and physical organic chemistry into the fascinating realm of pyrotechnics and explosives. The field of high-energy chemistry lost an important leader when Dr. McLain passed away in 1981.

John A. Conkling

CONTENTS

CHEMISTRY OF PYROTECHNICS

Fireworks burst in the sky over the Washington Monument to celebrate Independence Day. Such fireworks combine all of the effects that can be created using pyrotechnic mixtures. A fuse made with black powder provides a time delay between lighting and launching. A propellant charge—also black powder—lifts each fireworks cannister hundreds of feet into the air. There, a "bursting charge" ruptures the casing while igniting numerous small "stars"--pellets of composition that burn with vividly-colored flames. (Zambelli Internationale)

1 INTRODUCTION

This book is an introduction to the basic principles and theory of pyrotechnics. Much of the material is also applicable to the closely-related areas of propellants and explosives. The term "high-energy chemistry" will be used to refer to all of these fields. Explosives rapidly release large amounts of energy, and engineers take advantage of this energy and the associated shock and pressure to do work. Pyrotechnic mixtures react more slowly, producing light, color, smoke, heat, noise, and motion.

The chemical reactions involved are of the electron-transfer, or oxidation-reduction, type. The compounds and mixtures to be studied are almost always solids and are designed to function in the absence of external oxygen. The reaction rates to be dealt with range along a continuum from very slow burning to "instantaneous" detonations with rates greater than a kilometer per second (Table 1.1).

It is important to recognize early on that the same material may vary *dramatically* in its reactivity depending on its method of preparation and the conditions under which it is used. Black powder is an excellent example of this variability, and it is quite fitting that it serve as the first example of a "high-energy material" due to its historical significance. Black powder is an intimate mixture of potassium nitrate (75% by weight), charcoal (15%), and sulfur (10%). A *reactive* black powder is no simple material to prepare. If one merely mixes the three components briefly, a powder is produced that is difficult to light and burns quite slowly. The same ingredients in the same proportions, when thoroughly mixed, moistened, and ground with a heavy stone wheel to achieve a high degree of homogeneity, readily

TABLE 1.1 Classes of "High-Energy" Reactions

Class	Approximate reaction velocity	Example
Burning	Millimeters/second	Delay mixtures, colored smoke composition
Deflagration	Meters/second	Rocket propellants, confined black powder
Detonation	> 1 Kilometer/second	Dynamite, TNT

ignite and burn rapidly. Particle size, purity of starting materials, mixing time, and a variety of other factors are all critical in producing high-performance black powder. Also, deviations from the 75/15/10 ratio of ingredients will lead to substantial changes in performance. Much of the history of modern Europe is related to the availability of high-quality black powder for use in rifles and cannons. A good powder-maker was essential to military success, although he usually received far less recognition and decoration than the generals who relied upon his product.

The burning behavior of black powder illustrates how a pyrotechnic mixture can vary in performance depending on the conditions of its use. A small pile of loose black powder can be readily ignited by the flame from a match, producing an orange flash and a puff of smoke, but almost no noise. The same powder, sealed in a paper tube but still in loose condition, will explode upon ignition, rupturing the container with an audible noise. Black powder spread in a thin trail will quickly burn along the trail, a property used in making fuses. Finally, if the powder is compressed in a tube, one end is left open, and that end is then constricted to partially confine the hot gases produced when the powder is ignited, a rocket-type device is produced. This varied behavior is quite typical of pyrotechnic mixtures and illustrates why one must be quite specific in giving instructions for preparing and using the materials discussed in this book.

Why should someone working in pyrotechnics and related areas bother to study the basic chemistry involved? Throughout the 400-year "modern" history of the United States many black powder factories have been constructed and put into operation. Although smokeless powder and other new materials have replaced black powder as a propellant and delay mixture in many applications, there is still a sizeable demand for black powder in both the military and civilian pyrotechnic industries. How many black powder factories are operating in the United States today? Exactly *one*. The remainder have been destroyed by explosions or closed because of the probability of one occurring. In spite of a demand for the product, manufacturers are reluctant to engage in the production of the material because of the history of problems with accidental ignition during the manufacturing process. Why is black powder so sensitive to ignition? What can the chemist do to minimize the hazard? Can one alter the performance of black powder by varying the ingredients and their percentages, using theory as the approach rather than trial-and-error? It is this type of problem and its analysis that I hope can be addressed a bit more scientifically with an understanding of the fundamental concepts presented in this book. If one accident can be prevented as a consequence of someone's better insight into the chemical nature of high-energy materials, achieved through study of this book, then the effort that went into its preparation was worthwhile.

BRIEF HISTORY

The use of chemicals to produce heat, light, smoke, noise, and motion has existed for several thousand years, originating most likely in China or India. India has been cited as a particularly good possibility due to the natural deposits of saltpeter (potassium nitrate, KNO_3) found there [1].

Much of the early use of chemical energy involved military applications. "Greek fire," first reported in the 7th century A.D., was probably a blend of sulfur, organic fuels, and saltpeter that generated flames and dense fumes when ignited. It was used in a variety of incendiary ways in both sea and land battles and added a new dimension to military science [2].

At some early time, prior to 1000 A.D., an observant scientist recognized the unique properties of a blend of potassium nitrate, sulfur, and charcoal; and black powder was developed as the first "modern" high-energy composition. A formula quite similar to the one used today was reported by Marcus Graecus ("Mark the Greek") in an 8th century work "Book of Fires for Burning the Enemy" [2]. Greek fire and rocket-type devices were also discussed in these writings.

The Chinese were involved in pyrotechnics at an early date and had developed rockets by the 10th century [1]. Fireworks were available in China around 1200 A.D., when a Spring Festival reportedly used over 100 pyrotechnic sets, with accompanying music, blazing candle lights, and merriment. The cost of such a display was placed at several thousand liangs of silver (one liang = 31.2 grams) [3]. Chinese firecrackers became a popular item in the United States when trade was begun in the 1800's. Chinese fireworks remain popular in the United States today, accounting for well over half of the annual sales in this country. The Japanese also produce beautiful fireworks, but, curiously, they do not appear to have developed the necessary technology until fireworks were brought to Japan around 1600 A.D. by an English visitor [4]. Many of the advances in fireworks technology over the past several centuries have come from these two Asian nations.

The noted English scientist Roger Bacon was quite familiar with potassium nitrate/charcoal/sulfur mixtures in the 13th century, and writings attributed to him give a formula for preparing "thunder and lightning" composition [5]. The use of black powder as a propellant for cannons was widespread in Europe by the 14th century.

Good-quality black powder was being produced in Russia in the 15th century in large amounts, and Ivan the Terrible reportedly had 200 cannons in his army in 1563 [6]. Fireworks were being used for celebrations and entertainment in Russia in the 17th century, with Peter the First among the most enthusiastic supporters of this artistic use of pyrotechnic materials.

By the 16th century, black powder had been extensively studied in many European countries, and a published formula dating to Bruxelles in 1560 gave a 75.0/15.62/9.38 ratio of saltpeter/charcoal/sulfur that is virtually the same as the mixture used today [5]!

The use of pyrotechnic mixtures for military purposes in rifles, rockets, and cannons developed simultaneously with the civilian applications such as fireworks. Progress in both areas followed advances in modern chemistry, as new compounds were isolated and synthesized and became available to the pyrotechnician. Berthollet's discovery of potassium chlorate in the 1780's resulted in the ability to produce brilliant flame colors using pyrotechnic compositions, and color was added to the effects of sparks, noise, and motion previously available using potassium nitrate-based compositions. Chlorate-containing color-producing formulas were known by the 1830's in some pyrotechnicians' arsenals.

The harnessing of electricity led to the manufacturing of magnesium and aluminum metals by electrolysis in the latter part of the 19th century, and bright white sparks and white light could then be produced. Strontium, barium, and copper compounds capable of producing vivid red, green, and blue flames also became commercially available during the 19th century, and modern pyrotechnic technology really took off.

Simultaneously, the discovery of nitroglycerine in 1846 by Sobrero in Italy, and Nobel's work with dynamite, led to the development of a new generation of true high explosives that were far superior to black powder for many blasting and explosives applications. The development of modern smokeless powder – using nitrocellulose and nitroglycerine – led to the demise of black powder as the main propellant for guns of all types and sizes.

Although black powder has been replaced in most of its former uses by newer, better materials, it is important to recognize the important role it has played in modern civilization. Tenney Davis, addressing this issue in his classic book on the chemistry of explosives, wrote "The discovery that a mixture of potassium nitrate, charcoal, and sulfur is capable of doing useful work is one of the most important chemical discoveries or inventions of all time...the discovery of the controllable force of gunpowder, which made huge engineering achievements possible, gave access to coal and to minerals within the earth, and brought on directly the age of iron and steel and with it the era of machines and of rapid transportation and communication" [5].

Explosives are widely used today throughout the world for mining, excavation, demolition, and military purposes. Pyrotechnics are also widely used by the military for signalling and training. Civilian applications of pyrotechnics are many and varied, ranging from the common match to highway warning flares to the ever-popular fireworks.

The fireworks industry remains perhaps the most visible example of pyrotechnics, and also remains a major user of traditional black powder. This industry provides the pyrotechnician with the opportunity to fully display his skill at producing colors and other brilliant visual effects.

Fireworks form a unique part of the cultural heritage of many countries [7]. In the United States, fireworks have traditionally been associated with Independence Day – the 4th of July. In England, large quantities are set off in commemoration of Guy Fawkes Day (November 5th), while the French use fireworks extensively around Bastille Day (July 14th). In Germany, the use of fireworks by the public is limited to one hour per year – from midnight to 1 a.m. on January 1st, but it is reported to be quite a celebration. Much of the Chinese culture is associated with the use of firecrackers to celebrate New Year's and other important occasions, and this custom has carried over to Chinese communities throughout the world. The brilliant colors and booming noises of fireworks have a universal appeal to our basic senses.

To gain an understanding of how these beautiful effects are produced, we will begin with a review of some basic chemical principles and then proceed to discuss various pyrotechnic systems.

REFERENCES

1. U.S. Army Material Command, Engineering Design Handbook, Military Pyrotechnic Series, Part One, "Theory and Application," Washington, D.C., 1967 (AMC Pamphlet 706-185).
2. J. R. Partington, *A History of Greek Fire and Gunpowder*, W. Heffer and Sons Ltd., Cambridge, Eng., 1960.
3. Ding Jing, "Pyrotechnics in China," presented at the 7th International Pyrotechnics Seminar, Vail, Colorado, July, 1980.
4. T. Shimizu, *Fireworks — The Art, Science and Technique*, pub. by T. Shimizu, distrib. by Maruzen Co., Ltd., Tokyo, 1981.
5. T. L. Davis, *The Chemistry of Powder and Explosives*, John Wiley & Sons, Inc., New York, 1941.
6. A. A. Shidlovskiy, *Principles of Pyrotechnics*, 3rd Edition, Moscow, 1964. (Translated as Report FTD-HC-23-1704-74 by Foreign Technology Division, Wright-Patterson Air Force Base, Ohio, 1974.)
7. G. Plimpton, *Fireworks: A History and Celebration*, Doubleday, New York, 1984.

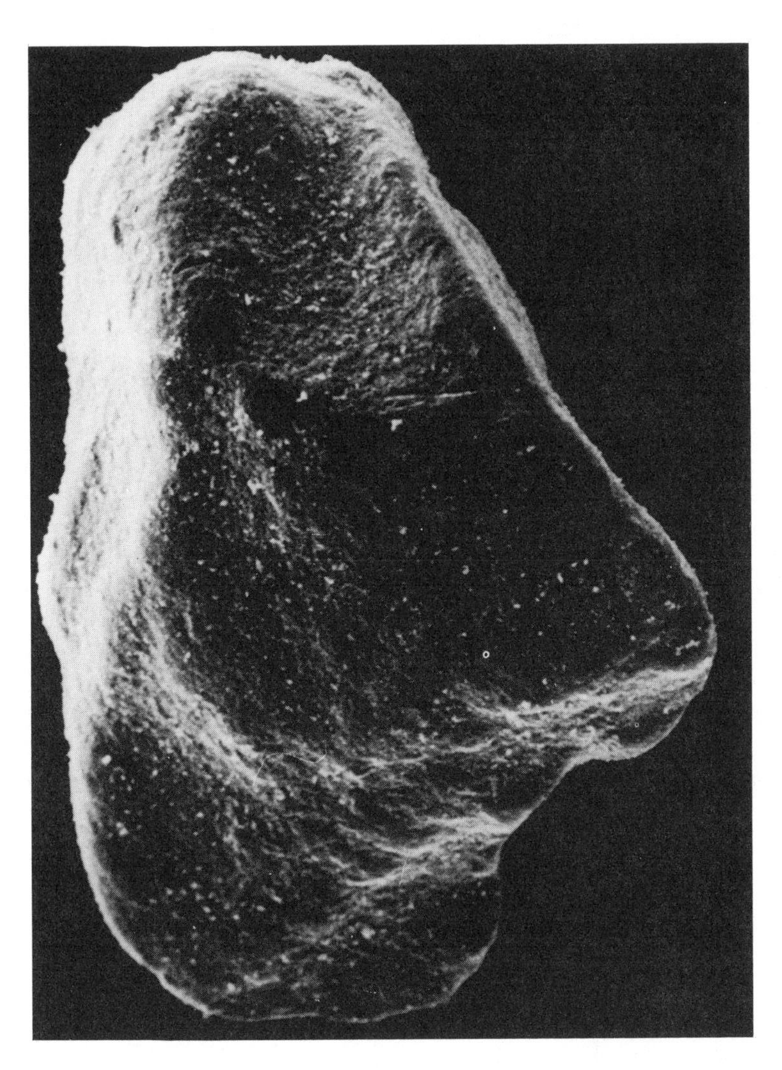

A grain of commercially-produced black powder, magnified 80 times. Extensive mixing and grinding of moist composition produces a homogeneous mixture of high reactivity. A mixture of the same three components—potassium nitrate, sulfur, and charcoal—that is prepared by briefly stirring the materials together will be much less reactive. (J. H. McLain files)

2
BASIC CHEMICAL PRINCIPLES

ATOMS AND MOLECULES

To understand the chemical nature of pyrotechnics, one must begin at the atomic level. Two hundred years of experiments and calculations have led to our present picture of the atom as the fundamental building block of matter.

An atom consists of a small, dense nucleus containing positively-charged *protons* and neutral *neutrons*, surrounded by a large cloud of light, negatively-charged *electrons*. Table 2.1 summarizes the properties of these subatomic particles.

A particular *element* is defined by its *atomic number* – the number of protons in the nucleus (which will equal the number of electrons surrounding the nucleus in a neutral atom). For example, iron is the element of atomic number 26, meaning that every iron atom will have 26 protons in its nucleus. Chemists use a one or two-letter symbol for each element to simplify communication; iron is given the symbol Fe, from the old Latin word for iron, ferrum. The sum of the protons plus neutrons found in a nucleus is called the mass number. For some elements only one mass number is found in nature. Fluorine (atomic number 9, mass number 19) is an example of such an element. Other elements are found in nature in more than one mass number. Iron is found as mass number 56 (91.52%), 54 (5.90%), 57 (2.245%), and 58 (0.33%). These different mass numbers of the same element are called *isotopes*, and vary in the number of neutrons found in the nucleus. *Atomic weight* refers to the average mass found in nature of all the atoms of a particular element; the atomic weight of iron is 55.847. For calculation purposes, these

TABLE 2.1 Properties of the Subatomic Particles

Particle	Location	Charge	Mass, amus[a]	Mass, grams
Proton	In nucleus	+1	1.007	1.673×10^{-24}
Neutron	In nucleus	0	1.009	1.675×10^{-24}
Electron	Outside nucleus	-1	0.00549	9.11×10^{-28}

[a]amu = atomic mass unit, where 1 amu = 1.66×10^{-24} gram.

atomic weights are used for the mass of a particular element. Table 2.2 contains symbols, atomic numbers, and atomic weights for the elements.

Chemical reactivity, and therefore pyrotechnic and explosive behavior, is determined primarily by the tendency for each element to gain or lose electrons during a chemical reaction. Calculations by theoretical chemists, with strong support from experimental studies, suggest that electrons in atoms are found in "orbitals," or regions in space where they possess the lowest possible energy — close to the nucleus but away from other negatively-charged electrons. As electrons are placed into an atom, energy levels close to the positive nucleus are occupied first, and the higher energy levels are then successively populated. Extra stability appears to be associated with completely filled levels, termed "shells." Elements with completely filled shells include helium (atomic number 2), neon (atomic number 10), argon (atomic number 18), and krypton (atomic number 36). These elements all belong to a group called the "inert gases," and their virtual lack of any chemical reactivity provides support for the theory of filled-shell stability.

Other elements show varying tendencies to obtain a filled shell by the sharing of electrons with other atoms, or by the actual gain or loss of electrons to form charged species, called *ions*. For example, sodium (symbol Na, atomic number 11) readily loses one electron to form the sodium ion, Na^+, with 10 electrons. By losing one electron, sodium has acquired the same number of electrons as the inert gas neon, and it has become a very stable chemical species. Fluorine (symbol F, atomic number 9) readily acquires one additional electron to become the fluoride ion, F^-. This is another 10-electron species and is

TABLE 2.2 Symbols, Atomic Weights, and Atomic Numbers of the Elements

Element	Symbol	Atomic number	Atomic weight, amus[a]
Actinium	Ac	89	-
Aluminum	Al	13	26.9815
Americium	Am	95	-
Antimony	Sb	51	121.75
Argon	Ar	18	39.948
Arsenic	As	33	74.9216
Astatine	At	85	-
Barium	Ba	56	137.34
Berkelium	Bk	97	-
Beryllium	Be	4	9.0122
Bismuth	Bi	83	208.980
Boron	B	5	10.811
Bromine	Br	35	79.909
Cadmium	Cd	48	112.40
Calcium	Ca	20	40.08
Californium	Cf	98	-
Carbon	C	6	12.01115
Cerium	Ce	58	140.12
Cesium	Cs	55	132.905
Chlorine	Cl	17	35.453
Chromium	Cr	24	51.996
Cobalt	Co	27	58.9332
Copper	Cu	29	63.54
Curium	Cm	96	-
Dysprosium	Dy	66	162.50
Einsteinium	Es	99	-
Erbium	Er	68	167.26
Europium	Eu	63	151.96
Fermium	Fm	100	-
Fluorine	F	9	18.9984
Francium	Fr	87	-
Gadolinium	Gd	64	157.25
Gallium	Ga	31	69.72
Germanium	Ge	32	72.59
Gold	Au	79	196.967
Hafnium	Hf	72	178.49
Helium	He	2	4.0026

TABLE 2.2 (continued)

Element	Symbol	Atomic number	Atomic weight, amus[a]
Holmium	Ho	67	164.930
Hydrogen	H	1	1.00797
Indium	In	49	114.82
Iodine	I	53	126.9044
Iridium	Ir	77	192.2
Iron	Fe	26	55.847
Krypton	Kr	36	83.80
Lanthanum	La	57	138.91
Lead	Pb	82	207.19
Lithium	Li	3	6.939
Lutetium	Lu	71	174.97
Magnesium	Mg	12	24.312
Manganese	Mn	25	54.9380
Mendelevium	Md	101	-
Mercury	Hg	80	200.59
Molybdenum	Mo	42	95.94
Neodymium	Nd	60	144.24
Neon	Ne	10	20.183
Neptunium	Np	93	-
Nickel	Ni	28	58.71
Niobium	Nb	41	92.906
Nitrogen	N	7	14.0067
Nobelium	No	102	-
Osmium	Os	76	190.2
Oxygen	O	8	15.9994
Palladium	Pd	46	106.4
Phosphorus	P	15	30.9738
Platinum	Pt	78	195.09
Plutonium	Pu	94	-
Polonium	Po	84	-
Potassium	K	19	39.102
Praseodymium	Pr	59	140.907
Promethium	Pm	61	-
Protactinium	Pa	91	-
Radium	Ra	88	-
Radon	Rn	86	-
Rhenium	Re	75	186.2
Rhodium	Rh	45	102.905

TABLE 2.2 (continued)

Element	Symbol	Atomic number	Atomic weight, amus[a]
Rubidium	Rb	37	85.47
Ruthenium	Ru	44	101.07
Samarium	Sm	62	150.35
Scandium	Sc	21	44.956
Selenium	Se	34	78.96
Silicon	Si	14	28.086
Silver	Ag	47	107.870
Sodium	Na	11	22.9898
Strontium	Sr	38	87.62
Sulfur	S	16	32.064
Tantalum	Ta	73	180.948
Technetium	Tc	43	-
Tellurium	Te	52	127.60
Terbium	Tb	65	158.924
Thallium	Tl	81	204.37
Thorium	Th	90	232.038
Thulium	Tm	69	168.934
Tin	Sn	50	118.69
Titanium	Ti	22	47.90
Tungsten	W	74	183.85
Uranium	U	92	238.03
Vanadium	V	23	50.942
Xenon	Xe	54	131.30
Ytterbium	Yb	70	173.04
Yttrium	Y	39	88.905
Zinc	Zn	30	65.37
Zirconium	Zr	40	91.22

[a]amu = atomic mass unit, where 1 amu = 1.66×10^{-24} gram.

quite stable. Other elements display similar tendencies to gain or lose electrons to acquire "inert gas" electron configurations by becoming positive or negative ions. Many chemical species found in nature are *ionic compounds*. These are crystalline solids composed of interpenetrating lattices of positive and negative ions held together by electrostatic attraction between these oppositely-charged particles. Table salt, or sodium chloride, is an ionic compound consisting of sodium and chloride ions, Na^+ and Cl^-, and one uses the formula NaCl to represent the one-to-one ionic ratio. The attractive forces holding the solid together are called *ionic bonds*.

Hence, if one brings together a good electron donor (such as a sodium atom) and a good electron acceptor (such as a fluorine atom), one might expect a chemical reaction to occur. Electrons are transferred and an ionic compound (sodium fluoride, NaF) is produced.

$Na + F \rightarrow Na^+F^-$ (sodium fluoride)

A three-dimensional solid lattice of sodium and fluoride ions is created, where each sodium ion is surrounded by fluoride ions, and each fluoride ion in turn is surrounded by sodium ions. Another very important aspect of such a reaction is the fact that *energy* is released as the product is formed. This release of energy associated with product formation is most important in the consideration of the chemistry of pyrotechnics.

In addition to forming ions by electron transfer, atoms may *share* electrons with other atoms as a means of acquiring filled shells (and their associated stability). The simplest illustration of this is the combination of two hydrogen atoms (symbol H, atomic number 1) to form a hydrogen *molecule*.

$H + H \rightarrow H\text{-}H$ (H_2, a molecule)

The sharing of electrons between two atoms is called a *covalent bond*. Such bonds owe their stability to the interaction of the shared electrons with *both* positive nuclei. The nuclei will be separated by a certain distance – termed the *bond distance* – that maximizes the nuclear-electron attractions balanced against the nuclear-nuclear repulsion. A *molecule* is a neutral species of two or more atoms held together by covalent bonds.

The element *carbon* (symbol C) is almost always found in nature covalently bonded to other carbon atoms or to a variety of other elements (most commonly H, O, and N). Due to the presence of carbon-containing compounds in all living things, the chemistry of carbon compounds is known as *organic chemistry*. Most high explosives are organic compounds. TNT (trinitrotoluene), for example, consists of C, H, N, and O atoms, with a molecular formula of $C_7H_5N_3O_6$. We will encounter other organic compounds in our study of fuels and binders in pyrotechnic mixtures.

Covalent bonds can form between dissimilar elements, such as hydrogen and chlorine.

TABLE 2.3 Electronegativity Values for Some Common Elements

Element	Pauling electronegativity value[a]
Fluorine, F	4.0
Oxygen, O	3.5
Nitrogen, N	3.0
Chlorine, Cl	3.0
Bromine, Br	2.8
Carbon, C	2.5
Sulfur, S	2.5
Iodine, I	2.5
Phosphorus, P	2.1
Hydrogen, H	2.1

[a]*Source*: L. Pauling, *The Nature of the Chemical Bond*, Cornell University Press, Ithaca, NY, 1960.

$$H + Cl \rightarrow H\text{-}Cl \text{ (hydrogen chloride)}$$

By this combination, both atoms now have "filled shell" electronic configurations and a hydrogen chloride molecule is formed. The sharing here is not exactly equal, however, for chlorine is a stronger electron attractor than hydrogen. The chlorine end of the molecule is slightly electron rich; the hydrogen end is electron deficient. This behavior can be noted using the Greek letter "delta" as the symbol for "partial," as in

$$^{\delta+}H\text{-}Cl^{\delta-}$$

The bond that is formed in hydrogen chloride is termed *polar covalent*, and a molecule possessing these partial charges is referred to as "polar." The relative ability of atoms of different elements to attract electron density is indicated by the property termed *electronegativity*. A scale ranking the elements was developed by Nobel Laureate Linus Pauling. The electronegativity sequence for some of the more common covalent-bond forming elements is given in Table 2.3. Using this sequence, one can assign partial charges to atoms in a variety of molecules; the more electronegative atom in a given bond will bear the partial negative charge, leaving the other atom with a partial positive charge.

TABLE 2.4 Boiling Points of Several Small Molecules

Compound	Formula	Boiling point (°C at 1 atmosphere pressure)
Methane	CH_4	-164
Carbon dioxide	CO_2	-78.6
Hydrogen sulfide	H_2S	-60.7
Water	H_2O	+100

$^{\delta-}$F-H$^{\delta+}$ $^{\delta-}$N-I$^{\delta+}$ $^{\delta-}$O-C$^{\delta+}$

These partial charges, or *dipoles*, can lead to intermolecular attractions that play an important role in such physical properties as melting point and boiling point, and they are quite important in determining solubility as well. The boiling point of water, 100°C, is quite high when compared to values for other small molecules (Table 2.4).

This high boiling point for water can be attributed to strong intermolecular attractions (called "dipole-dipole interactions") of the type

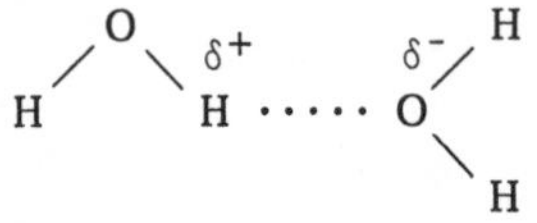

H–O–H$^{\delta+}$ ····· $^{\delta-}$O(H)–H

The considerable solubility of polar molecules and many ionic compounds in water can be explained by dipole-dipole or ion-dipole interactions between the dissolved species and the solvent, water.

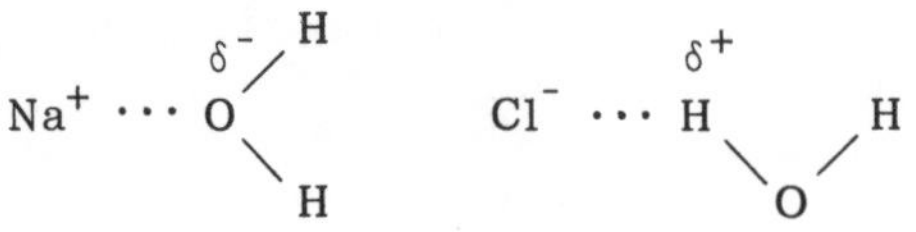

Na^+ ··· $^{\delta-}$O(H)–H Cl^- ··· $^{\delta+}$H–O–H

The solubility of solid compounds in water, as well as in other solvents, is determined by the competition between attractions in the solid state between molecules or ions and the solute-solvent attractions that occur in solution. A solid that is more attracted to itself than to solvent molecules will not dissolve. A general rule of solubility is "likes dissolve likes" – a polar solvent such as water is most effective at dissolving polar molecules (such as sugar) and ionic compounds. A non-polar solvent such as gasoline is most effective at dissolving other non-polar species such as motor oil, but it is a poor solvent for ionic species such as sodium chloride or potassium nitrate.

THE MOLE CONCEPT

Out of the atomic theory developed by John Dalton and other chemistry pioneers in the 19th century grew a number of important concepts essential to an understanding of all areas of chemistry, including pyrotechnics and explosives. The basic features of the atomic theory are:

1. The atom is the fundamental building block of matter, and consists of a collection of positive, negative, and neutral subatomic particles.
   Approximately 90 naturally-occurring elements are known to exist (additional elements have recently been synthesized in the laboratory using nuclear reactions, but these unstable species are not found in nature).
2. Elements may combine to form more complex species called *compounds*. The *molecule* is the fundamental unit of a compound and consists of two or more atoms joined together by chemical bonds.
3. All atoms of the same element are identical in terms of the number of protons and electrons contained in the neutral species. Atoms of the same element may vary in the number of neutrons, and therefore may vary in mass.
4. The chemical reactivity of an atom depends on the number of electrons; therefore, the reactivity of all atoms of a given element should be the same, and reproducible, anywhere in the world.
5. Chemical reactions consist of the combination or recombination of atoms, in fixed ratios, to produce new species.
6. A relative scale of atomic weights (as the weighted average of all forms, or isotopes, of a particular element found in nature) has been developed. The base of this scale is the assignment of a mass of 12.0000 to the isotope of carbon containing 6 protons, 6 neutrons, and 6 electrons. An atomic weight table can be found in Table 2.2.

7. As electrons are placed into atoms, they successively occupy higher energy levels, or shells. Electrons in filled levels are unimportant as far as chemical reactivity is concerned. It is the outer, partially-filled level that determines chemical behavior. Hence, elements with the same outer-shell configuration display markedly similar chemical reactivity. This phenomenon is called *periodicity*, and an arrangement of the elements placing similar elements in a vertical column has been developed – the *periodic table*. The alkali metals (lithium, sodium, potassium, rubidium, and cesium) are one family of the periodic table – they all have one reactive electron in their outer shell. The halogens (fluorine, chlorine, bromine, and iodine) are another common family – all have seven electrons in their outer shell and readily accept an eighth electron to form a filled level.

The mass of one atom of any element is infinitessimal and is impossible to measure on any existing balance. A more convenient mass unit was needed for laboratory work, and the concept of the *mole* emerged, where one mole of an element is a quantity equal to the atomic weight in grams. One mole of carbon, for example, is 12.01 grams, and one mole of iron is 55.85 grams. The actual number of atoms in one mole of an element has been determined by several elegant experimental procedures to be 6.02×10^{23}! This quantity is known as *Avogadro's number*, in honor of one of the pioneers of the atomic theory. One can then see that one mole of carbon atoms (12.01 grams) will contain exactly the same number of atoms as one mole (55.85 grams) of iron. Using the mole concept, the chemist can now go into the laboratory and weigh out equal quantities of atoms of the various elements.

The same concept holds for molecules. One mole of water (H_2O) consists of 6.02×10^{23} molecules and has a mass of 18.0 grams. It contains one mole of oxygen atoms and two moles of hydrogen atoms covalently bonded to make water molecules. The *molecular weight* of a compound is the sum of the respective atomic weights, taking into account the number of atoms of each element that comprise the molecule. For ionic compounds, a similar concept termed *formula weight* is used. The formula weight of sodium nitrate, $NaNO_3$, is therefore:

$$\text{Na} + \text{N} + 3\ \text{O's} = 23.0 + 14.0 + 3(16.0) = 85.0\ \text{g/mole}$$

These concepts permit the chemist to examine chemical reactions and determine the mass relationships that are involved. For example, consider the simple pyrotechnic reaction

$KClO_4$	+ $4\ Mg$	$\rightarrow$ KCl	+ $4\ MgO$
1 mole	4 moles	1 mole	4 moles
138.6 g	97.2 g	74.6 g	161.2 g

In a balanced chemical equation, the number of atoms of each element on the left-hand, or reactant, side will equal the number of atoms of each element on the right-hand, or product, side. The above equation states that one mole of potassium perchlorate ($KClO_4$, a *reactant*) will react with 4 moles of magnesium metal to produce one mole of potassium chloride (KCl) and 4 moles of magnesium oxide (MgO).

In mass terms, 138.6 grams (or pounds, tons, etc.) of potassium perchlorate will react with 97.2 grams (or any other mass unit) of magnesium to produce 74.6 grams of KCl and 161.2 grams of MgO. This mass *ratio* will always be maintained regardless of the quantities of starting material involved. If 138.6 grams (1.00 mole) of $KClO_4$ and 48.6 grams (2.00 moles) of magnesium are mixed and ignited, only 69.3 grams (0.50 mole) of the $KClO_4$ will react, completely depleting the magnesium. Remaining as "excess" starting material will be 0.50 mole (69.3 grams) of $KClO_4$ – there is no magnesium left for it to react with! The products formed in this example would be 37.3 grams (0.50 mole) of KCl and 80.6 grams (2.00 moles) of MgO, plus the 69.3 grams of excess $KClO_4$.

The preceding example also illustrates the *law of conservation of mass*. In any normal chemical reaction (excluding nuclear reactions) the mass of the starting materials will always equal the mass of the products (including the mass of any excess reactant). 200 grams of a $KClO_4$/Mg mixture will produce 200 grams of products (which includes any excess starting material).

The "formula" for the preceding illustration involved $KClO_4$ and Mg in a 138.6 to 97.2 mass ratio. The balanced mixture – with neither material present in excess – should then be 58.8% $KClO_4$ and 41.2% Mg by weight. The study of chemical weight relationships of this type is referred to as *stoichiometry*. A mixture containing exactly the quantities of each starting material corresponding to the balanced chemical equation is referred to as a *stoichiometric mixture*. Such balanced compositions are frequently associated with maximum performance in high-energy chemistry and will be referred to in future chapters.

ELECTRON TRANSFER REACTIONS

Oxidation-Reduction Theory

A major class of chemical reactions involves the transfer of one or more electrons from one species to another. This process is referred to as an electron-transfer or *oxidation-reduction* reaction, where the species undergoing electron loss is said to be *oxidized* while the species acquiring electrons is *reduced*. Pyrotechnics, propellants, and explosives belong to this chemical reaction category.

The determination of whether or not a species has undergone a loss or gain of electrons during a chemical reaction can be made by assigning "oxidation numbers" to the atoms of the various reacting species and products, according to the following simple rules:

1. Except in a few rare cases, hydrogen is always +1 and oxygen is always -2. Metal hydrides and peroxides are the most common exceptions. (This rule is applied first — it has highest priority, and the rest are applied in decreasing priority.)
2. Simple ions have their charge as their oxidation number. For example, Na^+ is +1, Cl^- is -1, Al^{+3} is +3, etc. The oxidation number of an element in its standard state is 0.
3. In a polar covalent molecule, the more electronegative atom in a bonded pair is assigned all of the electrons shared between the two atoms. For example, in H-Cl, the chlorine atom is assigned both bonded electrons, making it identical to Cl^- and giving it an oxidation number of -1. The hydrogen atom therefore has an oxidation number of +1 (in agreement with rule #1 as well).
4. In a neutral molecule, the sum of the oxidation numbers will be 0. For an ion, the sum of the oxidation numbers on all the atoms will equal the net charge on the ion.

Examples

NH_3 (ammonia): The 3 hydrogen atoms are all +1 by rule 1. The nitrogen atom will therefore be -3 by rule 4.

CO_3^{-2} (the carbonate ion): The three oxygen atoms are all -2 by rule 1. Since the ion has a net charge of -2, the oxidation number of carbon will be $3(-2) + x = -2$, $x = +4$ by rule 4.

For the reaction

$KClO_4 + ?\ Mg \rightarrow KCl + ?\ MgO$

the oxidation numbers on the various atoms are:

$KClO_4$: This is an ionic compound, consisting of the potassium ion, K^+, and the perchlorate ion, ClO_4^-. The oxidation number of potassium in K^+ will be +1 by rule 2. In ClO_4^-, the 4 oxygen atoms are all -2, making the chlorine atom +7, by rule 4.

Mg: Magnesium is present in elemental form as a reactant, making its oxidation number 0 by rule 2.

KCl: This is an ionic compound made up of K^+ and Cl^- ions, with respective oxidation numbers of +1 and -1 by rule 2.

MgO: This is another ionic compound. Oxygen will be -2 by rule 1, leaving the magnesium ion as +2.

Examining the various changes in oxidation number that occur as the reaction proceeds, one can see that potassium and oxygen are unchanged going from reactants to products. Magnesium, however, undergoes a change from 0 to +2, corresponding to a loss of two electrons per atom — it has lost electrons, or been *oxidized*. Chlorine undergoes an oxidation number change from +7 to -1, or a *gain* of 8 electrons per atom — it has been *reduced*. In a balanced oxidation-reduction reaction, the electrons lost must equal the electrons gained; therefore, *four* magnesium atoms (each losing two electrons) are required to reduce one chlorine atom from the +7 (as ClO_4^-) to -1 (as Cl^-) state. The equation is now balanced!

$KClO_4 + 4\ Mg \rightarrow KCl + 4\ MgO$

Similarly, the equation for the reaction between potassium nitrate and sulfur can be balanced if one knows that the products are potassium oxide, sulfur dioxide, and nitrogen gas:

$?\ KNO_3 + ?\ S \rightarrow ?\ K_2O + ?\ N_2 + ?\ SO_2$

Again, analysis of the oxidation numbers reveals that potassium and oxygen are unchanged, with values of +1 and -2, respectively, on both sides of the equation. Nitrogen changes from a value of +5 in the nitrate ion (NO_3^-) to 0 in elemental form as N_2. Sulfur changes from 0 in elemental form to a value of +4 in SO_2. In this reaction, then, sulfur is oxidized and nitrogen is reduced. To balance the equation, 4 nitrogen atoms, each gaining 5 electrons, and 5 sulfur atoms, each losing 4 electrons, are required. This results in 20 electrons gained and 20 electrons lost – they're balanced. The balanced equation is therefore:

$$4\ KNO_3 + 5\ S \rightarrow 2\ K_2O + 2\ N_2 + 5\ SO_2$$

The ratio by weight of potassium nitrate and sulfur corresponding to a balanced – or stoichiometric – mixture will be 4(101.1) = 404.4 grams (4 moles) of KNO_3 and 5(32.1) = 160.5 grams (5 moles) of sulfur. This equals 72% KNO_3 and 28% S by weight. An ability to balance oxidation-reduction equations can be quite useful in working out weight ratios for optimum pyrotechnic performance.

Electrochemistry

If one takes a spontaneous electron-transfer reaction and *separates* the materials undergoing oxidation and reduction, allowing the electron transfer to occur through a good conductor such as a copper wire, a battery is created. By proper design, the electrical energy associated with reactions of this type can be harnessed. The fields of electrochemistry (e.g., batteries) and pyrotechnics (e.g., fireworks) are actually very close relatives. The reactions involved in the two areas can look strikingly similar:

$Ag_2O + Zn \rightarrow 2\ Ag + ZnO$ (a battery reaction)

$Fe_2O_3 + 2\ Al \rightarrow 2\ Fe + Al_2O_3$ (a pyrotechnic reaction)

In both fields of research, one is looking for inexpensive, high-energy electron donors and acceptors that will readily yield their energy on demand yet be quite stable in storage.

Electrochemists have developed extensive tables listing the relative tendencies of materials to donate or accept electrons, and these tables can be quite useful to the pyrochemist in his search for new materials. Chemicals are listed in order of decreasing tendency to gain electrons, and are all expressed as *half-reactions* in the reduction direction, with the half-reaction

$H^+ + e \rightarrow 1/2\ H_2$ 0.000 volts

arbitrarily assigned a value of 0.000 volts. All other species are measured relative to this reaction, with more readily-reducible species having *positive* voltages (also called standard reduction potentials), and less-readily reducible species showing negative values. Species with sizeable negative potentials should then, logically, be the best electron *donors*, and a combination of a good electron donor with a good electron acceptor should produce a battery of high voltage. Such a combination will also be a potential candidate for a pyrotechnic system. One must bear in mind, however, that most of the values listed in the electrochemistry tables are for reactions in *solution*, rather than for solids, so direct calculations can't be made for pyrotechnic systems. Some good ideas for candidate materials can be obtained, however.

A variety of materials of pyrotechnic interest, and their standard reduction potentials at 25°C are listed in Table 2.5. Note the large positive values associated with certain oxygen-rich negative ions, such as the chlorate ion (ClO_3^-), and the large negative values associated with certain active metals such as aluminum (Al).

THERMODYNAMICS

There are a vast number of possible reactions that the chemist working in the explosives and pyrotechnics fields can write between various electron donors (fuels) and electron acceptors (oxidizers). Whether a particular reaction will be of possible use depends on two major factors:

1. Whether or not the reaction is *spontaneous*, or will actually occur if the oxidizer and fuel are mixed together.
2. The *rate* at which the reaction will proceed, or the time required for complete reaction to occur.

Spontaneity is determined by a quantity known as the *free energy change*, ΔG. "Δ" is the symbol for the upper-case Greek letter "delta," and stands for "change in."

The thermodynamic requirement for a reaction to be spontaneous (at constant temperature and pressure) is that the products are of lower free energy than the reactants, or that ΔG — the change in free energy associated with the chemical reaction — be a negative value. Two quantities comprise the free energy of a system at a given temperature. The first is the *enthalpy*, or heat content, represented by the symbol H. The second is the *entropy*, represented by the symbol S, which may be viewed as the randomness or disorder of the system. The free energy of a system, G, is equal to H-TS, where T is the temperature of the system on the Kelvin, or absolute, scale. (To convert from Celsius to Kelvin temperature, add 273 degrees to the Celsius

TABLE 2.5 Standard Reduction Potentials

Half-reaction	Standard potential @25°C, in volts[a]
$3 N_2 + 2H^+ + 2e \rightarrow 2 HN_3$	-3.1
$Li^+ + e \rightarrow Li$	-3.045
$H_2BO_3^- + H_2O + 3e \rightarrow B + 4 OH^-$	-2.5
$Mg^{+2} + 2e \rightarrow Mg$	-2.375
$HPO_3^= + 2 H_2O + 3e \rightarrow P + 5 OH^-$	-1.71
$Al^{+3} + 3e \rightarrow Al$ (in dil. NaOH soln.)	-1.706
$TiO_2 + 4 H^+ + 4e \rightarrow Ti + 2 H_2O$	-0.86
$SiO_2 + 4 H^+ + 4e \rightarrow Si + 2 H_2O$	-0.84
$S + 2e \rightarrow S^=$	-0.508
$Bi_2O_3 + 3 H_2O + 6e \rightarrow 2 Bi + 6 OH^-$	-0.46
$WO_3 + 6 H^+ + 6e \rightarrow W + 3 H_2O$	-0.09
$Fe^{+3} + 3e \rightarrow Fe$	-0.036
$2 H^+ + 2e \rightarrow H_2$	0.000
$NO_3^- + H_2O + 2e \rightarrow NO_2^- + 2 OH^-$	+0.01
$H_2SO_3 + 4 H^+ + 4e \rightarrow S + 3 H_2O$	+0.45
$NO_3^- + 4 H^+ + 3e \rightarrow NO + 2 H_2O$	+0.96
$IO_3^- + 6 H^+ + 6e \rightarrow I^- + 3 H_2O$	+1.195
$HCrO_4^- + 7 H^+ + 3e \rightarrow Cr^{+3} + 4 H_2O$	+1.195
$ClO_4^- + 8 H^+ + 8e \rightarrow Cl^- + 4 H_2O$	+1.37
$BrO_3^- + 6 H^+ + 6e \rightarrow Br^- + 3 H_2O$	+1.44
$ClO_3^- + 6 H^+ + 6e \rightarrow Cl^- + 3 H_2O$	+1.45
$PbO_2 + 4 H^+ + 2e \rightarrow Pb^{+2} + 2 H_2O$	+1.46
$MnO_4^- + 8 H^+ + 5e \rightarrow Mn^{+2} + 4 H_2O$	+1.49

[a]Reference 1.

value.) The free energy *change* accompanying a chemical reaction at constant temperature is therefore given by

$$\Delta G = G(\text{products}) - G(\text{reactants}) = \Delta H - T\Delta S \qquad (2.1)$$

For a chemical reaction to be spontaneous, or energetically favorable, it is desirable that ΔH, or the enthalpy change, be a negative value, corresponding to the liberation of heat by the reaction. Any chemical process that liberates heat is termed *exothermic*, while a process that absorbs heat is called *endothermic*. ΔH values for many high-energy reactions have been experimentally determined as well as theoretically calculated. The typical units for ΔH, or *heat of reaction*, are calories/mole or calories/gram. The new International System of units calls for energy values to be given in *joules*, where one calorie = 4.184 joules. Most thermochemical data are found with the calorie as the unit, and it will be used in this book in most instances. Some typical ΔH values for pyrotechnics are given in Table 2.6. *Note*: 1 kcal = 1 kilocalorie = 1,000 calories.

It is also favorable to have the entropy change, ΔS, be a positive value, making the $-T\Delta S$ term in equation 2.1 a negative value. A positive value for ΔS corresponds to an *increase* in the randomness or disorder of the system when the reaction occurs. As a general rule, entropy follows the sequence:

$$S(\text{solid}) < S(\text{liquid}) \ll S(\text{gas})$$

Therefore, a process of the type *solids* $\rightarrow$ *gas* (common to many high-energy systems) is particularly favored by the change in entropy occurring upon reaction. Reactions that evolve heat and form gases from solid starting materials should be favored thermodynamically and fall in the "spontaneous" category. Chemical processes of this type will be discussed in subsequent chapters.

Heat of Reaction

It is possible to calculate a heat of reaction for a high-energy system by assuming what the reaction products will be and then using available thermodynamic tables of *heats of formation*. "Heat of formation" is the heat associated with the formation of a chemical compound from its constituent elements. For example, for the reaction

$$2\ Al + 3/2\ O_2 \rightarrow Al_2O_3$$

ΔH is -400.5 kcal/mole of Al_2O_3, and this value is therefore the heat of formation (ΔH_f) of aluminum oxide (Al_2O_3). The reaction

TABLE 2.6 Typical ΔH Values for "High-Energy" Reactions

Composition (% by weight)		ΔH (kcal/gram)[a]	Application
$KClO_4$	60	2.24	Photoflash
Mg	40		
$NaNO_3$	60	2.00	White light
Al	40		
Fe_2O_3	75	0.96	Thermite (heat)
Al	25		
KNO_3	75	0.66	Black powder
C	15		
S	10		
$KClO_3$	57	0.61	Red light
$SrCO_3$	25		
Shellac	18		
$KClO_3$	35	0.38	Red smoke
Lactose	25		
Red dye	40		

[a]Reference 2. All values represent heat *released* by the reaction.

of 2.0 moles (54.0 grams) of aluminum with oxygen gas (48.0 grams) to form Al_2O_3 (1.0 mole, 102.0 grams) will liberate 400.5 kcal of heat – a sizeable amount! Also, to *decompose* 102.0 grams of Al_2O_3 into 54.0 grams of aluminum metal and 48.0 grams of oxygen gas, one must put 400.5 kcal of heat *into* the system – an amount equal in magnitude but opposite in sign from the heat of formation. The heat of formation of any *element* in its standard state at 25°C will therefore be 0 using this system.

A chemical reaction can be considered to occur in two steps:

1. Decomposition of the starting materials into their constituent elements, followed by
2. Subsequent reaction of these elements to form the desired products.

The *net* heat change associated with the overall reaction can then be calculated from:

$$\Delta H(\text{reaction}) = \Sigma\Delta H_f(\text{products}) - \Sigma\Delta H_f(\text{reactants}) \qquad (2.2)$$

(where Σ = "the sum of")

This equation sums up the heats of formation of all of the products from a reaction, and then subtracts from that value the heat required to dissociate all of the starting materials into their elements. The difference between these two values is the *net* heat change, or heat of reaction. The heats of formation of a number of materials of interest to the high-energy chemist may be found in Table 2.7. All values given are for a reaction occurring at 25°C (298 K).

Example 1:

Consider the following reaction, balanced using the "oxidation numbers" method

Reaction	$KClO_4$ +	4 Mg →	KCl +	4 MgO
Grams	138.6	97.2	74.6	161.2
Heat of formation (kcal/mole x # of moles)	-103.4	4(0)	-104.4	4(-143.8)

ΔH(reaction) = $\Sigma\Delta H_f$(products) - $\Sigma\Delta H_f$(reactants)
= [-104.4 + 4(-143.8)] - [-103.4 + 4(0)]
= -576.2 kcal/mole $KClO_4$
= -2.44 kcal/gram of stoichiometric mixture (obtained by dividing -576.2 kcal by 138.6 + 97.2 = 235.8 grams of starting material).

Example 2:

Reaction	4 KNO_3	+ 5 C →	2 K_2O	+ 2 N_2	+ 5 CO_2
Grams	404.4	60	188.4	56	220
Heat of formation (kcal/mole x # of moles)	4(-118.2)	5(0)	2(-86.4)	2(0)	5(-94.1)

ΔH(reaction) = $\Sigma\Delta H_f$(products) - $\Sigma\Delta H_f$(reactants)
= [2(-86.4) + 0 + 5(-94.1)] - [4(-118.2) + 5(0)]
= -643.3 - (-472.8)
= -170.5 kcal/equation as written (4 moles KNO_3)
= -42.6 kcal/mole KNO_3
= -0.37 kcal/gram of stoichiometric mixture (-170.5 kcal per 464.4 grams)

TABLE 2.7 Standard Heats of Formation at 25°C

Compound	Formula	ΔHformation (kcal/mole)[a]
OXIDIZERS		
Ammonium nitrate	NH_4NO_3	-87.4
Ammonium perchlorate	NH_4ClO_4	-70.58
Barium chlorate (hydrate)	$Ba(ClO_3)_2 \cdot H_2O$	-184.4
Barium chromate	$BaCrO_4$	-345.6
Barium nitrate	$Ba(NO_3)_2$	-237.1
Barium peroxide	BaO_2	-151.6
Iron oxide	Fe_2O_3	-197.0
Iron oxide	Fe_3O_4	-267.3
Lead chromate	$PbCrO_4$	-217.7[b]
Lead oxide (red lead)	Pb_3O_4	-171.7
Lead peroxide	PbO_2	-66.3
Potassium chlorate	$KClO_3$	-95.1
Potassium nitrate	KNO_3	-118.2
Potassium perchlorate	$KClO_4$	-103.4
Sodium nitrate	$NaNO_3$	-111.8
Strontium nitrate	$Sr(NO_3)_2$	-233.8
FUELS		
Elements		
Aluminum	Al	0
Boron	B	0
Iron	Fe	0
Magnesium	Mg	0
Phosphorus (red)	P	-4.2
Silicon	Si	0
Titanium	Ti	0
Organic Compounds[c]		
Lactose (hydrate)	$C_{12}H_{22}O_{11} \cdot H_2O$	-651
Shellac	$C_{16}H_{24}O_5$	-227
Hexachloroethane	C_2Cl_6	-54
Starch (polymer)	$(C_6H_{10}O_5)n$	-227 (per unit)
Anthracene	$C_{14}H_{10}$	+32
Polyvinyl chloride (PVC)	$(-CH_2CHCl-)_n$	-23 (per unit)[b]

TABLE 2.7 (continued)

Compound	Formula	$\Delta H_{formation}$ (kcal/mole)[a]
REACTION PRODUCTS		
Aluminum oxide	Al_2O_3	-400.5
Barium oxide	BaO	-133.4
Boron oxide	B_2O_3	-304.2
Carbon dioxide	CO_2	-94.1
Carbon monoxide	CO	-26.4
Chromium oxide	Cr_2O_3	-272.4
Lead oxide (Litharge)	PbO	-51.5
Magnesium oxide	MgO	-143.8
Nitrogen	N_2	0
Phosphoric acid	H_3PO_4	-305.7
Potassium carbonate	K_2CO_3	-275.1
Potassium chloride	KCl	-104.4
Potassium oxide	K_2O	-86.4
Potassium sulfide	K_2S	-91.0
Silicon dioxide	SiO_2	-215.9
Sodium chloride	NaCl	-98.3
Sodium oxide	Na_2O	-99.0
Strontium oxide	SrO	-141.5
Titanium dioxide	TiO_2	-225
Water	H_2O	-68.3
Zinc chloride	$ZnCl_2$	-99.2

[a]Reference 1.
[b]Reference 4.
[c]Reference 2.

RATES OF CHEMICAL REACTIONS

The preceding section discussed how the chemist can make a thermodynamic determination of the spontaneity of a chemical reaction. However, even if these calculations indicate that a reaction should be quite spontaneous (the value for ΔG is a large, negative number), there is no guarantee that the reaction will proceed rapidly when the reactants are mixed together at 25°C (298 K). For example, the reaction:

Wood + $O_2 \rightarrow CO_2 + H_2O$

has a large, negative value for ΔG at 25°C. However (fortunately!) wood and oxygen are reasonably stable when mixed together at 25°C (a typical room temperature). The explanation of this thermodynamic mystery lies in another energy concept known as the *energy of activation*. This term represents that amount of energy needed to take the starting materials from their reasonably stable form at 25°C and convert them to a reactive, higher-energy excited state. In this excited state, a reaction will occur to form the anticipated products, with the liberation of considerable energy – all that was required to reach the excited state, plus more. Figure 2.1 illustrates this process.

The rate of a chemical reaction is determined by the magnitude of this required activation energy, and rate is a temperature-dependent phenomenon. As the temperature of a system is raised, an exponentially-greater number of molecules will possess the necessary energy of activation. The reaction rate will therefore increase accordingly in an exponential fashion as the temperature rises. This is illustrated in Figure 2.2. Much of the pioneering work in the area of reaction rates was done by the Swedish chemist Svante Arrhenius, and the equation describing this rate-temperature relationship is known as the Arrhenius Equation:

$$k = Ae^{-E_a/RT} \qquad (2.3)$$

where

k = the rate constant for a particular reaction at temperature T. (This is a constant representing the speed of the reaction, and is determined experimentally.)

A = a temperature-independent constant for the particular reaction, termed the "pre-exponential factor."

E_a = the activation energy for the reaction.

R = a universal constant known as the "ideal gas constant."

T = temperature, in degrees Kelvin.

If the natural logarithm (ln) of both sides of equation (2.3) is taken, one obtains:

$$\ln k = \ln A - Ea/RT \qquad (2.4)$$

Therefore, if the rate constant, k, is measured at several temperatures and ln k versus 1/T is plotted, a straight line should

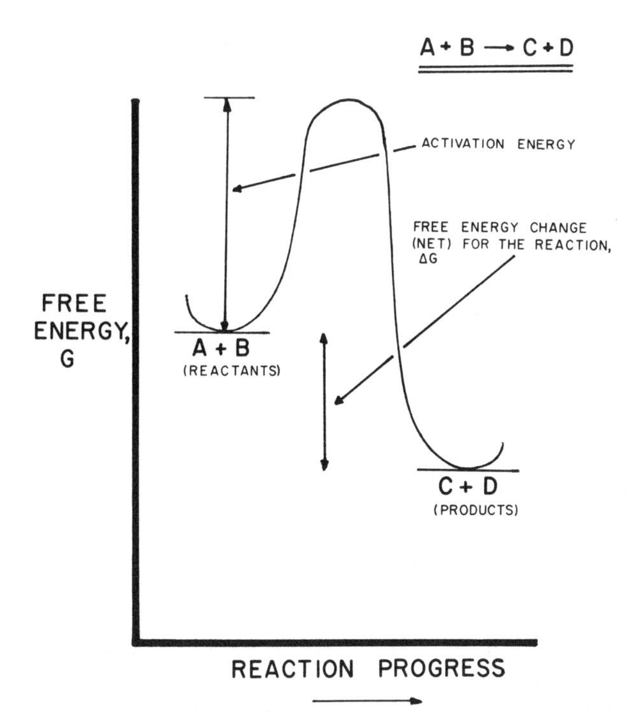

FIG. 2.1 The free energy, G, of a chemical system as reactants A and B convert to products C and D. A and B must first acquire sufficient energy ("activation energy") to be in a reactive state. As products C and D are formed, energy is released and the final energy level is reached. The *net* energy change, ΔG, corresponds to the difference between the energies of the products and reactants. The *rate* at which a reaction proceeds is determined by the energy barrier that must be crossed – the activation energy.

be obtained, with slope of $-Ea/R$. Activation energies can be obtained for chemical reactions through such experiments. The Arrhenius Equation, describing the rate-temperature relationship, is of considerable significance in the ignition of pyrotechnics and explosives, and it will be referred to in subsequent chapters.

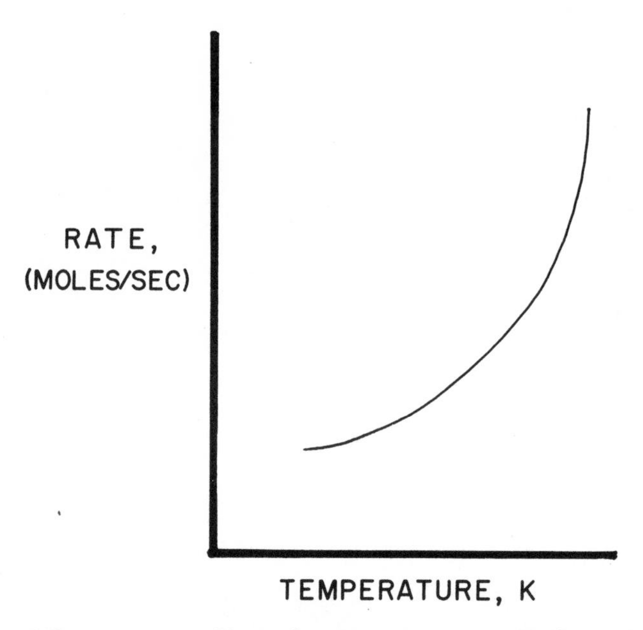

FIG. 2.2 The effect of temperature on reaction rate. As the temperature of a chemical system is increased, the rate at which that system reacts to form products increases exponentially.

ENERGY-RICH BONDS

Certain covalent chemical bonds (such as N-O and Cl-O) are particularly common in the high-energy field. Bonds between two highly electronegative atoms tend to be less stable than ones between atoms of differing electronegativity. The intense competition for the electron density in a bond such as Cl-O is believed to be responsible for at least some of this instability. A modern chemical bonding theory known as the "molecular orbital theory" predicts inherent instability for some common high-energy species. The azide ion, N_3^-, and the fulminate ion, CNO^-, are examples of species whose unstable behavior is explainable using this approach [3].

In structures such as the nitrate ion, NO_3^-, and the perchlorate ion, ClO_4^-, a highly electronegative atom has a large, positive oxidation number (+5 for N in NO_3^-, +7 for Cl in ClO_4^-). Such

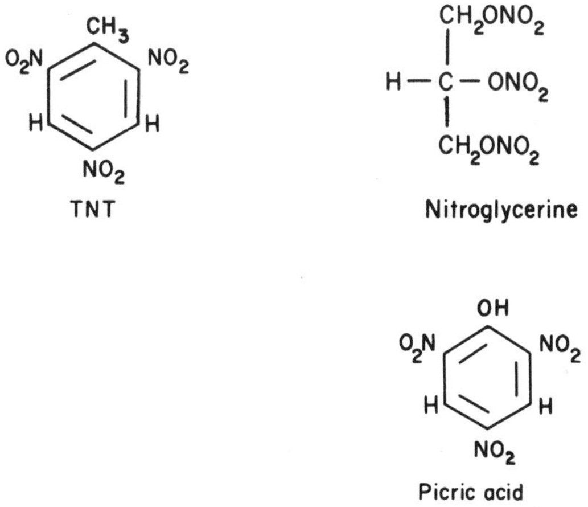

TNT

Nitroglycerine

Picric acid

FIG. 2.3 Many "unstable" organic compounds are used as explosives. These molecules contain internal oxygen, usually bonded to nitrogen, and undergo intramolecular oxidation-reduction to form stable products – carbon dioxide, nitrogen, and water. The "mixing" of oxidizer and fuel is achieved at the molecular level, and *fast* rates of decomposition can be obtained.

large positive numbers indicate electron deficiency. It is therefore not surprising that structures with such bonding arrangements are particularly reactive as electron acceptors (oxidizers). It is for similar reasons that many of the nitrated carbon-containing ("organic") compounds, such as nitroglycerine and TNT, are so unstable (Figure 2.3). The nitrogen atoms in these molecules want to accept electrons to relieve bonding stress, and the carbon atoms found in the same molecules are excellent electron donors. Two very stable gaseous (high entropy) chemical species, N_2 and CO_2, are produced upon decomposition of most nitrated carbon-containing compounds, helping to insure a large, negative value for ΔG for the decomposition (therefore making it a spontaneous process).

These considerations make it mandatory that anyone working with nitrogen-rich carbon-containing compounds or with nitrate, perchlorate, and similar oxygen-rich negative ions must use extreme caution in the handling of these materials until their properties have been fully examined in the laboratory. Elevated temperatures should also be avoided when working with potentially-unstable materials, because of the rate-temperature relationship that is exponential in nature. A non-existent or sluggish process can become an explosion when the temperature of the system is sharply increased.

STATES OF MATTER

With few exceptions the high-energy chemist deals with materials that are in the solid state at normal room temperature. Solids mix very slowly with one another, and hence they tend to be quite sluggish in their reactivity. Rapid reactivity is usually associated with the formation, at higher temperatures, of liquids or gases. Species in these states can diffuse into one another more rapidly, leading to accelerated reactivity.

In pyrotechnics, the solid-to-liquid transition appears to be of considerable importance in initiating a self-propagating reaction. The oxidizing agent is frequently the key component in such mixtures, and a ranking of common oxidizers by increasing melting point bears a striking resemblance to the reactivity sequence for these materials (Table 2.8).

Gases

On continued heating, a pure material passes from the solid to liquid to vapor state, with continued absorption of heat. The volume occupied by the vapor state is much greater than that of the solid and liquid phases. One mole (18 grams) of water occupies approximately 18 milliliters (0.018 liters) as a solid or liquid. One mole of water vapor, however, at 100°C (373 K) occupies approximately 30.6 liters at normal atmospheric pressure. The volume occupied by a gas can be estimated using the *ideal gas equation* (equation 2.5).

$$V = nRT/P \qquad (2.5)$$

where
V = volume occupied by the gas, in liters
n = # moles of gas
R = a constant, 0.0821 liter-atm/deg-mole
T = temperature, in K
P = pressure, in atmospheres

TABLE 2.8 Melting Points of Some Common Oxidizers

Oxidizer	Formula	Melting point, °C[a]
Potassium nitrate	KNO_3	334
Potassium chlorate	$KClO_3$	356
Barium nitrate	$Ba(NO_3)_2$	592
Potassium perchlorate	$KClO_4$	610
Strontium nitrate	$Sr(NO_3)_2$	570
Lead chromate	$PbCrO_4$	844
Iron oxide	Fe_2O_3	1565

[a]Reference 1.

This equation is obeyed quite well by the inert gases (helium, neon, etc.) and by small diatomic molecules such as H_2 and N_2. Molecules possessing polar covalent bonds tend to have strong intermolecular attractions and usually deviate from "ideal" behavior. Equation 2 remains a fairly good estimate of volume and pressure even for these polar molecules, however. Using the ideal gas equation, one can readily estimate the pressure produced during ignition of a confined high-energy composition.

For example, assume that 200 milligrams (0.200 grams) of black powder is confined in a volume of 0.1 milliliter. Black powder burns to produce approximately 50% gaseous products and 50% solids. Approximately 1.2 moles of permanent gas are produced per 100 grams of powder burned (the gases are mainly N_2, CO_2, and CO) [5]. Therefore, 0.200 grams should produce 0.0024 moles of gas, at a temperature near 2000 K. The expected pressure is:

$$P = \frac{(0.0024\ \text{mole})(0.0821\ \text{liter-atm/deg-mole})(2000\ \text{deg})}{(0.0001\ \text{liter})}$$

$$= 3941\ \text{atm!}$$

Needless to say, the casing will rupture and an explosion will be observed. Burning a similar quantity of black powder in the open, where little pressure accumulation occurs, will produce a slower, less violent (but still quite vigorous!) reaction and no explosive effect. This dependence of burning behavior on degree of confinement is an important characteristic of pyrotechnic mixtures, and distinguishes them from true high explosives.

Liquids

Gas molecules are widely separated, travelling at high speeds while colliding with other gas molecules and with the walls of their container. Pressure is produced by these collisions with the walls and depends upon the number of gas molecules present as well as their kinetic energy. Their speed, and therefore their kinetic energy, increases with increasing temperature.

As the temperature of a gas system is lowered, the speed of the molecules decreases. When these lower-speed molecules collide with one another, attractive forces between the molecules become more significant, and a temperature will be reached where *condensation* occurs – the vapor state converts to liquid. Dipole-dipole attractive forces are most important in causing condensation, and molecules with substantial partial charges, resulting from polar covalent bonds, typically have high condensation temperatures. (Condensation temperature will be the same as the boiling point of a liquid, approached from the opposite direction.)

The liquid state has a minimum of order, and the molecules have considerable freedom of motion. A drop of food coloring placed in water demonstrates the rapid diffusion that can occur in the liquid state. The solid state will exhibit no detectable diffusion. If this experiment is tried with a material such as iron, the liquid food coloring will merely form a drop on the surface of the metal.

At the liquid surface, molecules can acquire high vibrational and translational energy from their neighbors, and one will occasionally break loose to enter the vapor state. This phenomenon of vapor above a liquid surface is termed *vapor pressure*, and will lead to gradual evaporation of a liquid unless the container is covered. In this case, an *equilibrium* is established between the molecules entering the vapor state per minute and the molecules recondensing on the liquid surface. The pressure of gas molecules above a confined liquid is a constant for a given material at a given temperature, and is known as the *equilibrium vapor pressure*. It increases exponentially with increasing temperature. When the vapor pressure of a liquid is equal to the external pressure acting on the liquid surface, *boiling* occurs. For solids and liquids to undergo sustained burning, the presence of a portion of the fuel in the vapor state is required.

The Solid State

The solid state is characterized by definite shape and volume. The observed shape will be the one that maximizes favorable interactions between the atoms, ions, or molecules making up the structure. The preferred shape begins at the atomic or molecular level and is regularly repeated throughout the solid, producing a highly-symmetrical, three-dimensional form called a *crystal*. The network produced is termed the *crystalline lattice*.

Solids lacking an ordered, crystalline arrangement are termed *amorphous* materials, and resemble rigid liquids in structure and properties. Glass (SiO_2) is the classic example of an amorphous solid. Such materials typically soften on heating, rather than showing a sharp melting point.

In the crystalline solid state, there is little vibrational or translational freedom, and hence diffusion into a crystalline lattice is slow and difficult. As the temperature of a solid is raised by the input of heat, vibrational and translational motion increases. At a particular temperature – termed the *melting point* – this motion overcomes the attractive forces holding the lattice together and the liquid state is produced. The liquid state, on cooling, returns to the solid state as crystallization occurs and heat is released by the formation of strong attractive forces.

The types of solids, categorized according to the particles that make up the crystalline lattice, are listed in Table 2.9.

The type of crystalline lattice formed by a solid material depends on the size and shape of the lattice units, as well as on the nature of the attractive forces. Six basic crystalline systems are possible [6]:

1. Cubic: three axes of equal length, intersecting at all right angles
2. Tetragonal: three axes intersecting at right angles; only two axes are equal in length
3. Hexagonal: three axes of equal length in a single plane intersecting at 60° angles; a fourth axis of different length is perpendicular to the plane of the other three

TABLE 2.9 Types of Crystalline Solids

Type of solid	Units comprising crystal lattice	Attractive force	Examples
Ionic	Positive and negative ions	Electrostatic attraction	KNO_3, NaCl
Molecular	Neutral molecules	Dipole-dipole attractions, plus weaker, non-polar forces	CO_2 ("dry ice"), sugar
Covalent	Atoms	Covalent bonds	Diamond (carbon)
Metallic	Metal atoms	Dispersed electrons attracted to numerous metal atom nuclei	Fe, Al, Mg

4. Rhombic: three axes of unequal length, intersecting at right angles
5. Monoclinic: three axes of unequal length, two of which intersect at right angles
6. Triclinic: three axes of unequal length, none of which intersect at right angles

To this point, our model of the solid state has suggested a placement of every lattice object at the proper site to create a "perfect" three-dimensional crystal. Research into the actual structure of solids has shown that crystals are far from perfect, containing a variety of types of defects. Even the purest crystals modern chemistry can create contain large numbers of impurities and "misplaced" ions, molecules, or atoms in the lattice. These inherent defects can play an important role in the reactivity of solids by providing a mechanism for the transport of electrons and heat through the lattice. They also can greatly enhance the ability of another substance to diffuse into the lattice, thereby again affecting reactivity [7].

A commonly-observed phenomenon associated with the presence of impurities in a crystalline lattice is a depression in the

TABLE 2.10 Thermal Conductivity Values for Solids[a]

Material	Thermal conductivity (X 10^3), cal/sec-cm-°C
Copper	910
Aluminum	500
Iron	150
Glass	2.3
Oak wood	0.4
Paper	0.3
Charcoal	0.2

[a]Reference 8.

melting point of the solid, with the solid → liquid transition occurring over a broad range rather than displaying the sharp melting observed with a purer material. Melting behavior thereby provides a convenient means of checking the purity of solids.

An important factor in the ignition and propagation of burning of pyrotechnic compositions is the conduction of heat along a column of the mixture. Hot gases serve as excellent heat carriers, but frequently the heat must be conducted by the solid state, ahead of the reaction zone. Heat can be transferred by molecular motion as well as by free, mobile electrons [6]. The thermal conductivity values of some common materials are given in Table 2.10. Examining this table, one can readily see how the presence of a small quantity of metal powder in a pyrotechnic composition can greatly increase the thermal conductivity of the mixture, and thereby increase the burning rate.

Electrical conductivity can also be an important consideration in pyrotechnic theory [7]. This phenomenon results from the presence of mobile electrons in the solid that migrate when an electrical potential is applied across the material. Metals are the best electrical conductors, while ionic and molecular solids are generally much poorer, serving well as insulators.

ACIDS AND BASES

An *acid* is commonly defined as a molecule or ion that can serve as a hydrogen ion (H^+) donor. The hydrogen ion is identical to the proton – it contains one proton in the nucleus, and has no electrons surrounding the nucleus. H^+ is a light, mobile, reactive species. A *base* is a species that functions as a hydrogen ion acceptor. The transfer of a hydrogen ion (proton) from a good donor to a good acceptor is called an acid/base reaction. Materials that are neither acidic nor basic in nature are said to be *neutral*.

Hydrogen chloride (HCl) is a gas that readily dissolves in water. In water, HCl is called hydrochloric acid and the HCl molecule serves as a good proton donor, readily undergoing the reaction

$$HCl \rightarrow H^+ + Cl^-$$

to produce a hydrogen ion and a chloride ion in solution. The concentration of hydrogen ions in water can be measured by a variety of methods and provides a measure of the acidity of an aqueous system. The most common measure of acidity is pH, a number representing the negative common logarithm of the hydrogen ion concentration:

$$pH = -\log [H^+]$$

If a solution also contains hydroxide ion (OH^-), a good proton acceptor, the reaction

$$H^+ + OH^- \rightarrow H_2O$$

occurs, forming water – a neutral species. The overall reaction is represented by an equation such as

$$HCl + NaOH \rightarrow H_2O + NaCl$$

Acids usually contain a bond between hydrogen and an electronegative element such as F, O, or Cl. The electronegative element pulls electron density away from the hydrogen atom, giving it partial positive character and making it willing to leave as H^+. The presence of additional F, O, and Cl atoms in the molecule further enhances the acidity of the species. Examples of strong acids include sulfuric acid (H_2SO_4), hydrochloric acid (HCl), perchloric acid ($HClO_4$), and nitric acid (HNO_3).

Most of the common bases are ionic compounds consisting of a positive metal ion and the negatively-charged hydroxide ion, OH^-. Examples include sodium hydroxide (NaOH), potassium hydroxide

(KOH), and calcium hydroxide, $Ca(OH)_2$. Ammonia (NH_3) is a weak base, capable of reacting with H^+ to form the ammonium ion, NH_4^+.

Acids catalyze a variety of chemical reactions, even when present in small quantity. The presence of trace amounts of acidic materials in many high-energy compounds and mixtures can lead to instability. The chlorate ion, ClO_3^-, is notoriously unstable in the presence of strong acids. Chlorate-containing mixtures will usually *ignite* if a drop of concentrated sulfuric acid is added.

Many metals are also vulnerable to acids, undergoing an oxidation/reduction reaction that produces the metal ion and hydrogen gas. The balanced equation for the reaction between HCl and magnesium is

$$Mg + 2\ HCl \rightarrow Mg^{+2} + H_2 + 2\ Cl^- + \text{heat}$$

Consequently, most metal-containing compositions must be free of acidic impurities or extensive decomposition (and possibly ignition) may occur.

As protection against acidic impurities, high-energy mixtures will frequently contain a small percentage of a neutralizer. Sodium bicarbonate ($NaHCO_3$) and magnesium carbonate ($MgCO_3$) are two frequently-used materials. The carbonate ion, CO_3^{-2}, reacts with H^+

$$2\ H^+ + CO_3^{-2} \rightarrow H_2O + CO_2$$

to form two neutral species – water and carbon dioxide.

Boric acid (H_3BO_3) – a solid material that is a weak H^+ donor – is sometimes used as a neutralizer for base-sensitive compositions. Mixtures containing aluminum metal and a nitrate salt are notably sensitive to excess hydroxide ion, and a small percentage of boric acid can be quite effective in stabilizing such compositions.

INSTRUMENTAL ANALYSIS

Modern instrumental methods of analysis have provided scientists with a wealth of information regarding the nature of the solid state and the reactivity of solids. Knowledge of the structure of solids and an ability to study thermal behavior are essential to an understanding of the behavior of high-energy materials.

X-ray crystallography has provided the crystal type and lattice dimensions for numerous solids. In this technique, high-energy x-rays strike the crystal and are diffracted in a pattern characteristic of the particular lattice type. Complex mathematical analysis can convert the diffraction pattern into the actual crystal structure. Advances in computer technology have revolutionized this field in the past few years. Complex structures, formerly requiring months or years to determine, can now be analyzed in short order. Even huge protein and nucleic acid chains can be worked out by the crystallographer [9].

Differential thermal analysis (DTA) has provided a wealth of information regarding the thermal behavior of pure solids as well as solid mixtures [10]. Melting points, boiling points, transitions from one crystalline form to another, and decomposition temperatures can be obtained for pure materials. Reaction temperatures can be determined for mixtures, such as ignition temperatures for pyrotechnic and explosive compositions.

Differential thermal analysis detects the absorption or release of heat by a sample as it is heated at a constant rate from room temperature to an upper limit, commonly 500°C. Any heat-absorbing changes occurring in the sample (e.g., melting or boiling) will be detected, as will processes that evolve heat (e.g., exothermic reactions). These changes are detected by continually comparing the temperature of the sample with that of a thermally-inert reference material (frequently aluminum oxide) that undergoes no phase changes or reactions over the temperature range being studied. Both sample and reference are placed in glass capillary tubes, a thermocouple is inserted in each, and the tubes are placed in a metal heating block. Current is applied to the electric heater to produce a linear temperature increase (typically 20-50 degrees/minute) [7].

If an endothermic (heat-absorbing) process occurs, the sample will momentarily become cooler than the reference material; the small temperature difference is detected by the pair of thermocouples and a downward deflection, termed an *endotherm*, is produced in the plot of ΔT (temperature difference between sample and reference) versus T (temperature of the heating block). Evolution of heat by the sample will similarly produce an upward deflection, termed an *exotherm*. The printed output produced by the instrument, a *thermogram*, is a thermal "fingerprint" of the material being analyzed. Thermal analysis is quite useful for determining the purity of materials; this is accomplished by examining the location and "sharpness" of the melting point. DTA is also useful for qualitative identification of solid materials, by comparing the thermal pattern with those of known materials. Reaction temperatures, including the ignition temperatures of high-energy materials, can be quickly (and safely) measured by thermal analysis. These temperatures will correspond to conditions

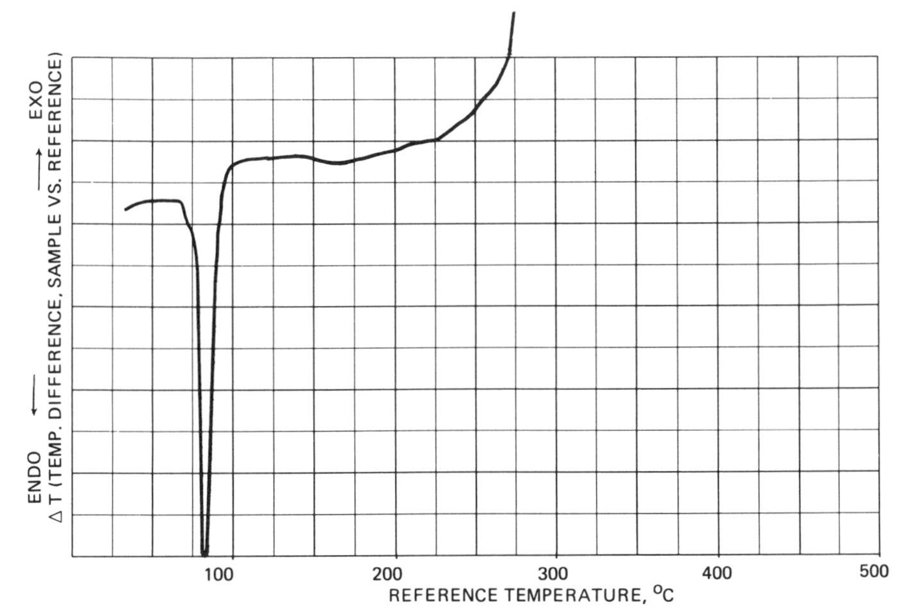

FIG. 2.4 The thermogram for pure 2,4,6-trinitrotoluene (TNT). The major features are an endotherm corresponding to melting at 81°C and an exothermic decomposition peak beginning near 280°. The x axis represents the temperature of the heating block in degrees centigrade. The y axis indicates the difference in temperature, ΔT, between the sample and an identically-heated reference solid, typically glass beads or aluminum oxide.

of rapid heating of a confined sample, and must be recognized as such.

Some representative thermograms of high-energy materials are shown in Figures 2.4-2.6.

LIGHT EMISSION

The pyrotechnic phenomena of heat, smoke, noise, and motion are reasonably easy to comprehend. Heat results from the rapid release of energy associated with the formation of stable chemical bonds during a chemical reaction. Smoke is produced by the dispersion in air of many small particles during a chemical reaction.

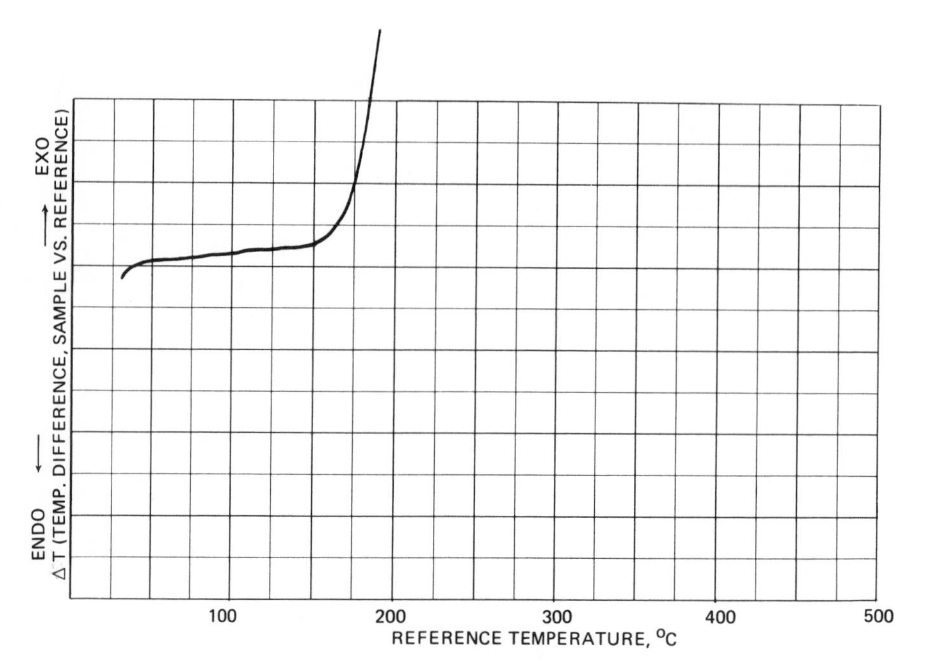

FIG. 2.5 Ballistite, a "smokeless powder" consisting of 60% nitrocellulose and 40% nitroglycerine, produces a thermogram with no transitions detectable prior to exothermic decomposition above 150°C.

Noise is produced by the rapid generation of gas at high temperature, creating waves that travel through air at the "speed of sound," 340 meters/second. Motion can be produced if you direct the hot gaseous products of a pyrotechnic reaction out through an exit, or nozzle. The thrust that is produced can move an object of considerable mass, if sufficient propellant is used.

The theory of color and light production, however, involves the energy levels available for electrons in atoms and molecules, according to the beliefs of modern chemical theory. In an atom or molecule, there are a number of "orbitals" or energy levels that an electron may occupy. Each of these levels corresponds to a discrete energy value, and *only* these energies are possible. The energy is said to be *quantized*, or restricted to certain values that depend on the nature of the particular atom or molecule.

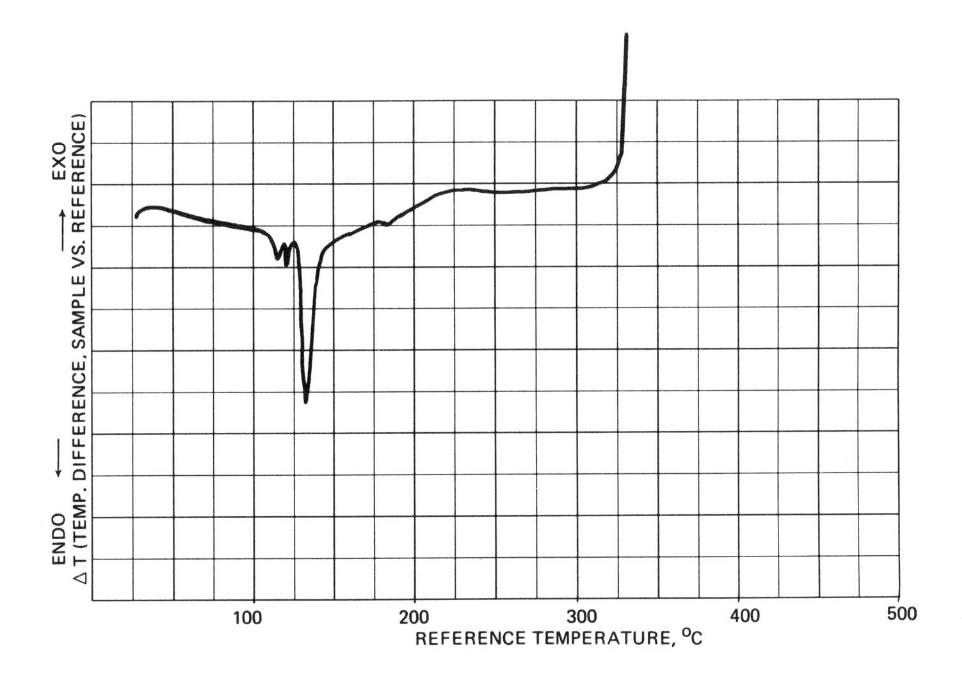

FIG. 2.6 Black powder was the first "modern" high-energy mixture, and it is still used in a variety of pyrotechnic applications. It is an intimate blend of potassium nitrate (75%), charcoal (15%), and sulfur (10%). The thermogram for the mixture shows endotherms near 105° and 119°C corresponding to a solid-solid phase transition and melting for sulfur, a strong endotherm near 130° representing a solid-solid transition in potassium nitrate, and a violent exotherm near 330°C where ignition of the mixture occurs.

We can represent these allowed electronic energies by a diagram such as Figure 2.7.

Logic suggests that an electron will occupy the lowest energy level available, and electrons will successively fill these levels as they are added to an atom or molecule. "Quantum mechanics" restricts all orbitals to a maximum of two electrons (these two have opposite "spins" and do not strongly repel one another), and hence a filling process occurs. The filling pattern for the sodium atom (sodium is atomic number 11 – therefore there will be 11 electrons in the neutral atom) is shown in Figure 2.7).

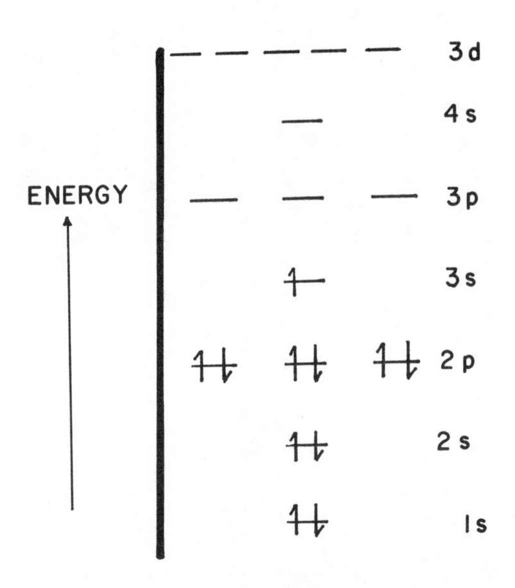

FIG. 2.7 The energy levels of the sodium atom. The sodium atom contains 11 electrons. These electrons will successively fill the lowest available energy levels in the atom, with a maximum population of two electrons in any given "orbital." The experimentally-determined energy level sequence is shown in this figure, with the 11th (and highest-energy) electron placed in the 3s level. The lowest vacant level is a 3p orbital. To raise an electron from the 3s to the 3p level requires 3.38×10^{-19} joules of energy. This energy corresponds to light of 589 nanometer wavelength – the yellow portion of the visible spectrum. Sodium atoms heated to high temperature will *emit* this yellow light as electrons are thermally excited to the 3p level, and then return to the 3s level and give off the excess energy as yellow light.

When energy is put into a sodium atom, in the form of heat or light, one means of accepting this energy is for an electron to be "promoted" to a higher energy level. The electron in this "excited state" is unstable and will quickly return to the ground state with the release of an amount of energy exactly equal to the energy difference between the ground and excited states. For the sodium atom, the difference between the highest occupied and lowest unoccupied levels is 3.38×10^{-19} joules/atom.

This energy can be lost as heat upon return to the ground state, or it can be released as a unit, or "photon," of light.

Light, or electromagnetic radiation, has both wave and particle or unit character associated with its behavior. Wavelengths range from very short (10^{-12} meters) for the "gamma rays" that accompany nuclear decay to quite long (10 meters) for radio waves.

All light travels at the same speed in a vacuum, with a value of 3×10^8 meters/second – the "speed of light." This value can be used for the speed of light in air as well.

The wavelength of light can now be related to the *frequency*, or number of waves passing a given point per second, using the speed of light value:

$$\text{frequency } (\nu) = \text{speed (c)/wavelength } (\lambda) \tag{2.6}$$

$$\text{(waves/second)} = \text{(meters/second)/(meters/wave)}$$

The entire range of wavelengths comprising "light" is known as the *electromagnetic spectrum* (Figure 2.8).

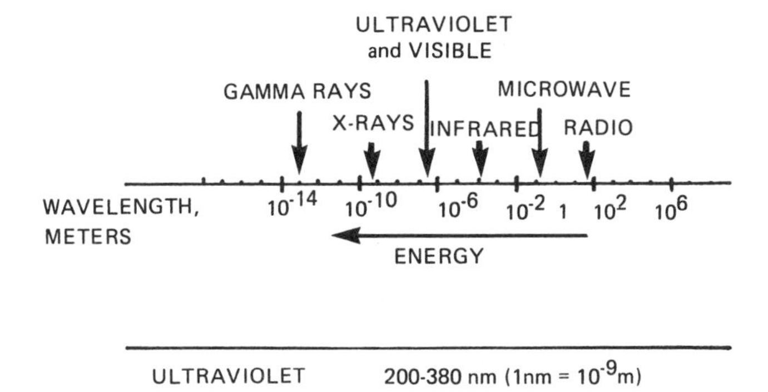

ULTRAVIOLET	200-380 nm (1nm = 10^{-9}m)
VISIBLE	380-780 nm

FIG. 2.8 The electromagnetic spectrum. The various regions of the electromagnetic spectrum correspond to a wide range of wavelengths, frequencies, and energies. The radiofrequency range is at the long-wavelength, low energy end, with gamma rays at the short-wavelength, high-frequency, high-energy end. The "visible" region – that portion of the spectrum perceived as color by the human visual system -- falls in the narrow region from 380-780 nanometers (1 nm = 10^{-9} m).

We can readily tell that light is a form of energy by staying out in the sun for too long a time. Elegant experiments by Einstein and others clearly showed that the energy associated with light was directly proportional to the *frequency* of the radiation:

$$E = h\nu = h\ c/\lambda \qquad (2.7)$$

where
E = the energy per light particle ("photon")
h = a constant, Planck's Constant, 6.63×10^{-34} joule-seconds
ν = frequency of light (in waves - "cycles" - per second)
c = speed of light (3×10^8 meters/second)
λ = wavelength of light (in meters)

This equation permits one to equate a wavelength of light with the energy associated with that particular radiation. For the sodium atom, the wavelength of light corresponding to the energy difference of 3.38×10^{-19} joules between the highest occupied and lowest unoccupied electronic energy levels should be:

$$E = h\nu = h\ c/\lambda$$

rearranging,

$$\lambda = h\ c/E$$

$$= \frac{(6.63 \times 10^{-34}\ \text{J-sec})(3 \times 10^{8}\ \text{m/sec})}{(3.38 \times 10^{-19}\ \text{J})}$$

$$= 5.89 \times 10^{-7}\ \text{meters}$$

$$= 589\ \text{nm (where 1 nm} = 10^{-9}\ \text{meters)}$$

Light of wavelength 589 nanometers falls in the yellow portion of the *visible* region of the electromagnetic spectrum. The characteristic yellow glow of sodium vapor lamps used to illuminate many highways results from this particular emission.

To produce this type of atomic emission in a pyrotechnic system, one must produce sufficient heat to generate atomic *vapor* in the flame, and then excite the atoms from the ground to various possible excited electronic states. Emission intensity will increase as the flame temperature increases, as more and more atoms are vaporized and excited. Return of the atoms to their ground state produces the light emission. A pattern of wavelengths, known as an *atomic spectrum*, is produced by each element. This pattern – a series of lines – corresponds to the various electronic transitions possible for the particular atom. The pattern is characteristic for each element and can be used for qualitative identification purposes.

Molecular Emission

A similar phenomenon is observed when molecules are vaporized and thermally excited. Electrons can be promoted from an occupied ground electronic state to a vacant excited state; when an electron returns to the ground state, a photon of light may be emitted.

Molecular spectra are usually more complex than atomic spectra. The energy levels are more complex, and vibrational and rotational sublevels superimpose their patterns on the electronic spectrum. *Bands* are generally observed rather than the sharp lines seen in atomic spectra. Emission intensity again increases as the flame temperature is raised. However, one must be concerned about reaching too high a temperature and decomposing the molecular emitter; the light emission pattern will change if this occurs. This is a particular problem in achieving an intense *blue* flame. The best blue light emitter – CuCl – is unstable at high temperature (above 1200°C).

"Black Body" Emission

The presence of solid particles in a pyrotechnic flame can lead to a substantial loss of color purity due to a complex process known as "black body radiation." Solid particles, heated to high temperature, radiate a continuous spectrum of light, much of it in the visible region – with the intensity exponentially increasing with temperature. If you are attempting to produce white light (which is a combination of *all* wavelengths in the visible region), this incandescent phenomenon is desirable.

Magnesium metal is found in most "white light" formulas. In an oxidizing flame, the metal is converted to the high-melting magnesium oxide, MgO, an excellent white-light emitter. Also, the high heat output of magnesium-containing compositions aids in achieving high flame temperatures. Aluminum metal is also commonly used for light production; other metals, including titanium and zirconium, are also good white-light sources.

The development of color and light-producing compositions will be considered in more detail in Chapter 7.

REFERENCES

1. R. C. Weast (Ed.), *CRC Handbook of Chemistry and Physics*, 63rd Ed., CRC Press, Inc., Boca Raton, Florida, 1982.
2. A. A. Shidlovskiy, *Principles of Pyrotechnics*, 3rd Edition, Moscow, 1964. (Translated as Report FTD-HC-23-1704-74 by Foreign Technology Division, Wright-Patterson Air Force Base, Ohio, 1974.)
3. L. Pytlewski, "The Unstable Chemistry of Nitrogen," presented at Pyrotechnics and Explosives Seminar P-81, Franklin Research Center, Philadelphia, Penna., August, 1981.
4. U.S. Army Material Command, Engineering Design Handbook, Military Pyrotechnic Series, Part Three, "Properties of Materials Used in Pyrotechnic Compositions," Washington, D.C., 1963 (AMC Pamphlet 706-187).
5. T. L. Davis, *The Chemistry of Powder and Explosives*, John Wiley & Sons, Inc., New York, 1941.
6. U.S. Army Material Command, Engineering Design Handbook, Military Pyrotechnic Series, Part One, "Theory and Application, Washington, D.C., 1967 (AMC Pamphlet 706-185).
7. J. H. McLain, *Pyrotechnics from the Viewpoint of Solid State Chemistry*, The Franklin Institute Press, Philadelphia, Penna., 1980.
8. R. L. Tuve, *Principles of Fire Protection Chemistry*, National Fire Protection Assn., Boston, Mass., 1976.
9. W. J. Moore, *Basic Physical Chemistry*, Prentice Hall, Englewood Cliffs, NJ, 1983.
10. W. W. Wendlandt, *Thermal Methods of Analysis*, Interscience, New York, 1964.

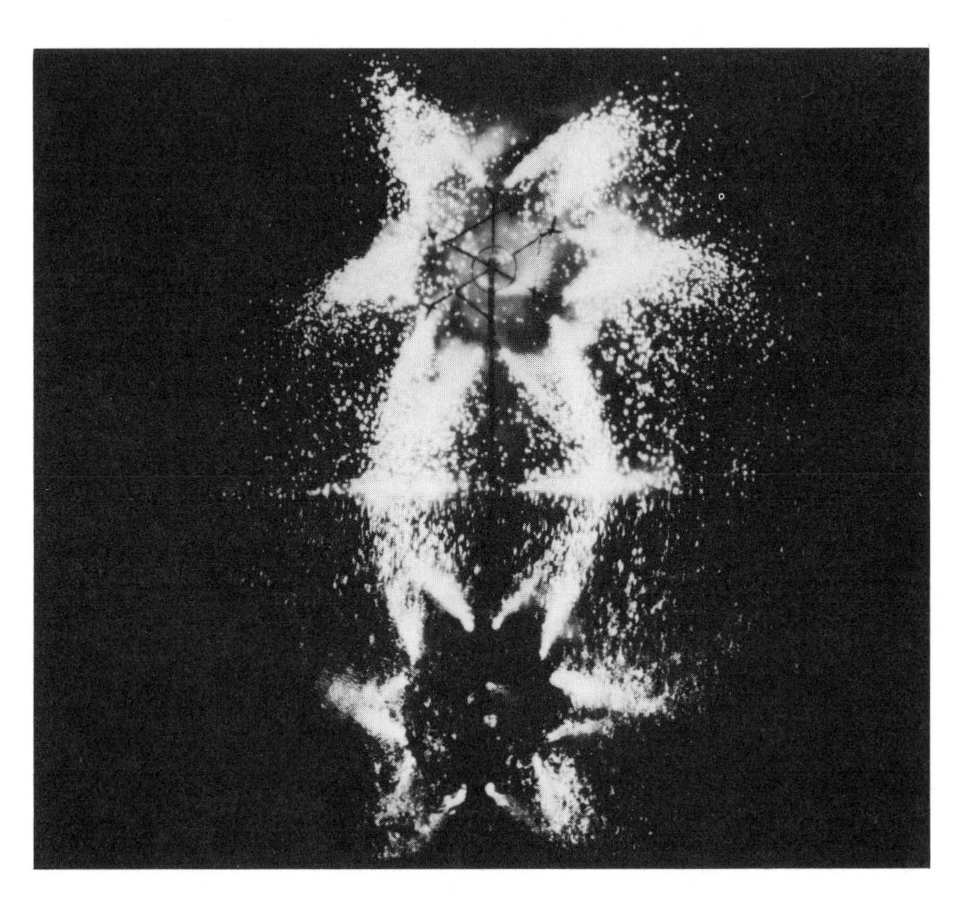

A "pinwheel" set piece, reflected over water. Cardboard tubes are loaded with spark-producing pyrotechnic composition. The "pinwheel," attached to a pole, revolves about its axis as hot gases are vented out the end of a "driver" tube to provide thrust. Sparks are produced by the burning of large particles of charcoal or aluminum. (Zambelli Internationale)

3
COMPONENTS OF HIGH-ENERGY MIXTURES

INTRODUCTION

Compounds containing both a readily-oxidizable and a readily-reducible component within one molecule are uncommon. Such species tend to have explosive properties. A molecule *or* ionic compound containing an internal oxidizer/reducer pair is inherently the most intimately-mixed high energy material that can be prepared. The mixing is achieved at the *molecular* (or *ionic*) level, and no migration or diffusion is required to bring the electron donor and electron acceptor together. The electron transfer reaction is *expected* to be rapid (even violent) in such species, upon application of the necessary activation energy to a small portion of the composition. A variety of compounds possessing this *intramolecular* reaction capability are shown in Table 3.1. The output from the exothermic decomposition of these compounds is typically heat, gas, and shock. Many of these materials *detonate* – a property quite uncommon with mixtures, where the degree of homogeneity is considerably less.

The high-energy chemist can greatly expand his repertoire of materials by preparing *mixtures*, combining an oxidizing material with a fuel to produce the exact heat output and burning rate needed for a particular application. Bright light, colors, and smoke can also be produced using such mixtures, adding additional dimensions to the uses of high-energy materials. For these effects to be achieved, it is critical that the mixture

TABLE 3.1 Compounds Containing Intramolecular Oxidation-Reduction Capability

Compound	Formula
Ammonium nitrate	NH_4NO_3
Ammonium perchlorate	NH_4ClO_4
Lead azide	$Pb(N_3)_2$
Trinitrotoluene (TNT)	$C_7H_5N_3O_6$
Nitroglycerine (NG)	$C_3H_5N_3O_9$
Mercury fulminate	$Hg(ONC)_2$

Note: These compounds readily undergo explosive decomposition when sufficient ignition stimulus is applied. A shock stimulus is frequently needed to activate the nonionic organic molecules (e.g., TNT); these compounds will frequently merely burn if a flame is applied.

burn rather than *explode*. Burning behavior is dependent upon a number of factors, and the pyrotechnist must carefully control these variables to obtain the desired performance.

Pyrotechnic mixtures "burn," but it must be remembered that these materials supply their *own* oxygen for combustion, through the thermal decomposition of an oxygen-rich material such as potassium chlorate:

$$2\ KClO_3 \xrightarrow{heat} 2\ KCl + 3\ O_2 \tag{3.1}$$

Thus, a pyrotechnic fire can *not* be suffocated – *no air* is needed for these mixtures to vigorously burn. In fact, confinement can *accelerate* the burning of a pyrotechnic composition by producing an increase in pressure, possibly leading to an explosion. Adequate venting is quite important in keeping a pyrotechnic fire from developing into a serious explosion.

A variety of ingredients, each serving one or more purposes, can be used to create an effective composition.

OXIDIZING AGENTS

Requirements

Oxidizing agents are usually oxygen-rich ionic solids that decompose at moderate-to-high temperatures, liberating oxygen gas. These materials must be readily available in pure form, in the proper particle size, at reasonable cost. They should give a neutral reaction when wet, be stable over a wide temperature range (at least up to 100°C), and yet readily decompose to release oxygen at higher temperatures. For the pyrotechnic chemist's use, acceptable species include a variety of negative ions (anions), usually containing high-energy Cl-O or N-O bonds:

NO_3^-	nitrate ion	ClO_3^-	chlorate ion
ClO_4^-	perchlorate ion	$CrO_4^=$	chromate ion
$O^=$	oxide ion	$Cr_2O_7^=$	dichromate ion

The positive ions used to combine with these anions must form compounds meeting several restrictions [1]:

1. The oxidizer must be quite low in *hygroscopicity*, or the tendency to acquire moisture from the atmosphere. Water can cause a variety of problems in pyrotechnic mixtures, and materials that readily pick up water may not be used. Sodium compounds in general are quite hygroscopic (e.g., sodium nitrate – $NaNO_3$) and thus they are rarely employed. Potassium salts tend to be much better, and are commonly used in pyrotechnics. Hygroscopicity tends to parallel water solubility, and solubility data can be used to anticipate possible moisture-attracting problems. The water solubility of the common oxidizers can be found in Table 3.2. However, it should be mentioned that large quantities of sodium nitrate are used by the military in combination with magnesium metal for white light production. Here, strict humidity control is required throughout the manufacturing process to avoid moisture uptake, and the finished items must be sealed to prevent water from being picked up during storage.
2. The oxidizer's positive ion (cation) must not adversely affect the desired flame color. Sodium, for example, is an intense emitter of yellow light, and its presence can ruin attempts to generate red, green, and blue flames.

TABLE 3.2 The Common Oxidizers and Their Properties

Compound	Formula	Formula weight	Melting point, °C[a]
Ammonium nitrate	NH_4NO_3	80.0	170
Ammonium perchlorate	NH_4ClO_4	117.5	Decomposes
Barium chlorate	$Ba(ClO_3)_2 \cdot H_2O$	322.3	414
Barium chromate	$BaCrO_4$	253.3	Decomposes
Barium nitrate	$Ba(NO_3)_2$	261.4	592
Barium peroxide	BaO_2	169.3	450
Iron oxide (red)	Fe_2O_3	159.7	1565
Iron oxide (black)	Fe_3O_4	231.6	1594
Lead chromate	$PbCrO_4$	323.2	844
Lead dioxide (lead peroxide)	PbO_2	239.2	290 (decomposes)
Lead oxide (litharge)	PbO	223.2	886
Lead tetroxide (red lead)	Pb_3O_4	685.6	500 (decomposes)
Potassium chlorate	$KClO_3$	122.6	356
Potassium nitrate	KNO_3	101.1	334
Potassium perchlorate	$KClO_4$	138.6	610
Sodium nitrate	$NaNO_3$	85.0	307
Strontium nitrate	$Sr(NO_3)_2$	211.6	570

[a]Reference 4.
[b]Reference 1.
[c]Reference 2.

Water solubility, grams/100 ml @ 20°C[a]	Heat of decomposition, kcal/mole	Heat of formation, kcal/mole[a]	Grams of oxygen released per gram of oxidizer	Weight of oxidizer required to evolve one gram of oxygen
118 (0°C)	-	-87.4	.60 (total 0)	-
37.2[c]	-	-70.6	Approx. 0.28	Approx. 3.5
27 (15°)	-28[b]	-184.4	.32	3.12
.0003 (16°)	-	-345.6	.095	10.6
8.7	+104[b]	-237.1	.31	3.27
Very slight	+17[b]	-151.6	.09	10.6
Insol.	-	-197.0	.30	3.33
Insol.	+266[b]	-267.3	.28	3.62
Insol.	-	-218	.074	13.5
Insol.	-	-66.3	.13 (total 0)	7.48
.0017	-	-51.5	.072 (total 0)	14.0
Insol.	-	-171.7	.093 (total 0)	10.7
7.1	-10.6[c]	-95.1	.39	2.55
31.6[c]	+75.5[b]	-118.2	.40	2.53
1.7[c]	-0.68[c]	-103.4	.46	2.17
92.1 (25°)[c]	+121[b]	-111.8	.47	2.13
70.9 (18°)	+92[c]	-233.8	.38	2.63

3. The alkali metals (Li, Na, K) and alkaline earth metals (Ca, Sr, and Ba) are preferred for the positive ion. These species are poor electron acceptors (and conversely, the metals are good electron donors), and they will not react with active metal fuels such as Mg and Al. If easily reducible metal ions such as lead (Pb^{+2}) and copper (Cu^{+2}) are present in oxidizers, there is a strong possibility that a reaction such as:

   $$Cu(NO_3)_2 + Mg \rightarrow Cu + Mg(NO_3)_2$$

   will occur, especially under moist conditions. The pyrotechnic performance will be greatly diminished, and spontaneous ignition might occur.
4. The compound must have an acceptable heat of decomposition. A value that is too exothermic will produce explosive or highly sensitive mixtures, while a value that is too endothermic will cause ignition difficulties as well as poor propagation of burning.
5. The compound should have as high an active oxygen content as possible. Light cations (Na^+, K^+, NH_4^+) are desirable while heavy cations (Pb^{+2}, Ba^{+2}) should be avoided if possible. Oxygen-rich anions, of course, are preferred.
6. Finally, all materials used in high-energy compositions should be low in toxicity, and yield low-toxicity reaction products.

In addition to ionic solids, covalent molecules containing halogen atoms (primarily F and Cl) can function as "oxidizers" in pyrotechnic compositions, especially with active metal fuels. Examples of this are the use of hexachloroethane (C_2Cl_6) with zinc metal in white smoke compositions,

$$3\ Zn + C_2Cl_6 \rightarrow 3\ ZnCl_2 + 2\ C$$

and the use of Teflon with magnesium metal in heat-producing mixtures,

$$(C_2F_4)_n + 2n\ Mg \rightarrow 2n\ C + 2n\ MgF_2 + \text{heat}$$

In both of these examples, the metal has been "oxidized" – has lost electrons and increased in oxidation number -- while the carbon atoms have gained electrons and been "reduced."

Table 3.2 lists some of the common oxidizers together with a variety of their properties.

Several oxidizers are so widely used that they merit special consideration. A few excellent books are available that provide additional details on the properties of these and other pyrotechnic materials [1, 2, 3].

Potassium Nitrate (KNO_3)

The oldest solid oxidizer used in high-energy mixtures, potassium nitrate (saltpeter) remains a widely-used ingredient well into the 20th century. Its advantages are ready availability at reasonable cost, low hygroscopicity, and the relative ease of ignition of many mixtures prepared using it. The ignitibility is related to the low (334°C) melting point of saltpeter. It has a high (39.6%) active oxygen content, decomposing at high temperature according to the equation

$$2\ KNO_3 \rightarrow K_2O + N_2 + 2.5\ O_2$$

This is a strongly *endothermic* reaction, with a ΔH value of +75.5 kcal/mole of KNO_3, meaning high energy-output fuels must be used with saltpeter to achieve rapid burning rates. When mixed with a simple organic fuel such as lactose, potassium nitrate may stop at the potassium nitrite (KNO_2) stage in its decomposition [2].

$$KNO_3 \rightarrow KNO_2 + 1/2\ O_2$$

With good fuels (charcoal or active metals), potassium nitrate will burn well. Its use in colored flame compositions is limited, primarily due to low reaction temperatures. Magnesium may be added to these mixtures to raise the temperature (and hence the light intensity), but the color value is diminished by "black body" emission from solid MgO.

Potassium nitrate has the additional property of not undergoing an explosion by itself, even when very strong initiating modes are used [2].

Potassium Chlorate ($KClO_3$)

One of the very best, and certainly the most controversial, of the common oxidizers is potassium chlorate, $KClO_3$. It is a white, crystalline material of low hygroscopicity, with 39.2% oxygen by weight. It is prepared by electrolysis from the chloride salt.

Potassium chlorate was used in the first successful colored-flame compositions in the mid-1800's and it remains in wide use today in colored smoke, firecrackers, toy pistol caps, matches, and color-producing fireworks.

However, potassium chlorate has been involved in a large percentage of the serious accidents at fireworks manufacturing plants, and it *must* be treated with great care if it is used at all. Other oxidizers are strongly recommended over this material, if one can be found that will produce the desired pyrotechnic effect.

Potassium chlorate compositions are quite prone to accidental ignition, especially if sulfur is also present. Chlorate/phosphorus mixtures are so reactive that they can *only* be worked with when quite wet. The high hazard of $KClO_3$ mixtures was gradually recognized in the late 19th century, and England *banned* all chlorate/sulfur compositions in 1894. United States factories have greatly reduced their use of potassium chlorate as well, replacing it with the less-sensitive potassium perchlorate in many formulas. The Chinese, however, continue to use potassium chlorate in firecracker and color compositions. Details on their safety record are not available, although several accidents are known to have occurred at their plants in recent years.

Several factors contribute to the instability of potassium chlorate-containing compositions. The first is the low (356°C) melting point and low decomposition temperature of the oxidizer. Soon after melting, $KClO_3$ decomposes according to equation 3.1.

$$2\ KClO_3 \rightarrow 2\ KCl + 3\ O_2 \qquad (3.1)$$

This reaction is quite vigorous, and becomes violent at temperatures above 500°C [2]. The actual decomposition mechanism may be more complex than equation 3.1 suggests. Intermediate formation of potassium perchlorate has been reported at temperatures just above the melting point, with the perchlorate then decomposing to yield potassium chloride and oxygen [5].

$$4\ KClO_3 \rightarrow 3\ KClO_4 + KCl$$

$$3\ KClO_4 \rightarrow 3\ KCl + 6\ O_2$$

$$\text{net:}\ 4\ KClO_3 \rightarrow 4\ KCl + 6\ O_2$$

The decomposition reaction of potassium chlorate is rare among the common oxidizers because it is *exothermic*, with a heat of reaction value of approximately -10.6 kcal/mole [2]. While most other oxidizers require a net heat *input* for their decomposition, potassium chlorate dissociates into KCl and O_2 with the liberation of heat. This heat output can lead to rate acceleration, and allows the ignition of potassium chlorate-containing compositions with a minimum of external energy input (ignition stimulus).

Potassium chlorate is particularly sensitive when mixed with sulfur, a low-melting (119°C) fuel. It is also sensitive when combined with low-melting organic compounds, and low ignition

TABLE 3.3 Ignition Temperatures of Potassium Chlorate/Fuel Mixtures

Fuel	Ignition temperature of stoichiometric mixture, C[a]
Lactose, $C_{12}H_{22}O_{11}$	195
Sulfur	220
Shellac	250
Charcoal	335
Magnesium powder	540
Aluminum powder	785
Graphite	890

[a]Reference 1.

temperatures are observed for most such compositions. Higher ignition temperatures are found for $KClO_3$/metal mixtures, attributable to the higher melting points and rigid crystalline lattices of these metallic fuels. However, these mixtures can be quite sensitive to ignition because of their substantial heat output, and should be regarded as quite hazardous. Ignition temperatures for some $KClO_3$ mixtures are given in Table 3.3. *Note:* Ignition temperatures are quite dependent upon the experimental conditions; a range of +/-50° may be observed, depending on sample size, heating rate, degree of confinement, etc. [6].

Mixtures containing potassium chlorate can be quite susceptible to the presence of a variety of chemical species. Acids can have a dramatic effect – the addition of a drop of concentrated sulfuric acid (H_2SO_4) to most $KClO_3$/fuel mixtures results in immediate inflammation of the composition. This dramatic reactivity has been attributed to the formation of chlorine dioxide (ClO_2) gas, a powerful oxidizer [5]. The presence of basic "neutralizers" such as magnesium carbonate and sodium bicarbonate in $KClO_3$ mixtures can greatly lower the sensitivity of these compositions to trace amounts of acidic impurities.

The ability of a variety of metal oxides -- most notably manganese dioxide, MnO_2 – to catalyze the thermal decomposition of potassium chlorate into potassium chloride and oxygen has been known for years. Little use is made of this behavior in pyrotechnics, however, because $KClO_3$ is almost *too* reactive in its normal state and ways are not needed to *enhance* its reactivity. Materials and methods to retard its decomposition are desired instead. However, knowledge of the ability of many materials to accelerate the decomposition of $KClO_3$ suggests that *impurities* could be quite an important factor in determining the reactivity and ignition temperature of chlorate-containing mixtures. It is vitally important that the $KClO_3$ used in pyrotechnic manufacturing operations be of the highest possible purity, and that all possible precautions be taken in storage and handling to prevent contamination of the material.

McLain has reported that potassium chlorate containing 2.8 mole% copper chlorate as an intentionally-added impurity (or "dopant") reacted *explosively* with sulfur at room temperature [7]! A pressed mixture of potassium chlorate with realgar (arsenic sulfide, As_2S_2) has also been reported to ignite at room temperature [2].

Ammonium chlorate, NH_4ClO_3, is an extremely unstable compound that decomposes violently at temperatures well below 100°C. If a mixture containing both potassium chlorate and an ammonium salt is prepared, there is a good possibility that an exchange reaction will occur – especially in the presence of moisture – to form some of the ammonium chlorate

$$NH_4X + KClO_3 \xrightarrow{H_2O} NH_4ClO_3 + KX$$

$$(X = Cl^-, NO_3^-, ClO_4^-, \text{etc.})$$

If this reaction occurs, the chance of spontaneous ignition of the mixture is likely. Therefore, any composition containing both a chlorate salt and an ammonium salt must be considered extremely hazardous. The shipping regulations of the United States Department of Transportation classify any such mixtures as "forbidden explosives" because of their instability [8]. However, compositions consisting of potassium chlorate, ammonium chloride, and organic fuels have been used, reportedly safely, for white smoke production [1].

Colored smoke compositions are a major user of potassium chlorate, and the safety record of these mixtures is excellent. A neutralizer (e.g., $MgCO_3$ or $NaHCO_3$) is typically added for storage stability, as well as to lower the reaction temperature through an endothermic decomposition, in the flame, of the type

$$MgCO_3 \xrightarrow{heat} MgO + CO_2$$

Colored smoke mixtures also contain either sulfur or a carbohydrate as the fuel, and a volatile organic dye that sublimes from the reaction mixture to produce the colored smoke. These compositions contain a large excess of potential fuel, and their explosive properties are greatly diminished as a result. Smoke mixtures *must* react with low flame temperatures (500°C or less) or the complex dye molecules will decompose, producing black soot instead of a brilliantly colored smoke. Potassium chlorate is far and away the best oxidizer for use in these compositions.

Potassium chlorate is truly a unique material. Shimizu has stated that no other oxidizer can surpass it for burning speed, ease of ignition, or noise production using a minimum quantity of composition [2]. It is also among the very best oxidizers for producing colored flames, with ammonium perchlorate as its closest rival. Chlorate-containing compositions can be prepared that will ignite and propagate at low flame temperatures – a property invaluable in colored smoke mixtures. By altering the fuel and the fuel/oxidizer ratio, much higher flame temperatures can be achieved for use in colored flame formulations. $KClO_3$ is a versatile material, but the inherent danger associated with it requires that alternate oxidizers be employed wherever possible. It is just *too* unstable and unpredictable to be safely used by the pyrotechnician in anything but colored smoke compositions, and even here coolants and considerable care are required!

Potassium Perchlorate ($KClO_4$)

This material has gradually replaced potassium chlorate ($KClO_3$) as the principal oxidizer in civilian pyrotechnics. Its safety record is far superior to that of potassium chlorate, although caution – including static protection – must still be used. Perchlorate mixtures, especially with a metal fuel such as aluminum, can have explosive properties, especially when present in bulk quantities and when confined.

Potassium perchlorate is a white, non-hygroscopic crystalline material with a melting point of 610°C, considerably higher than the 356°C melting point of $KClO_3$. It undergoes decomposition at high temperature

$$KClO_4 \xrightarrow{heat} KCl + 2\ O_2$$

forming potassium chloride and oxygen gas. This reaction has a slightly exothermic value of -0.68 kcal/mole [2] and produces substantial oxygen. The active oxygen content of $KClO_4$ – 46.2% – is one of the highest available to the pyrotechnician.

Because of its higher melting point and less-exothermic decomposition, potassium perchlorate produces mixtures that are less sensitive to heat, friction, and impact than those made with $KClO_3$ [2]. Potassium perchlorate can be used to produce colored flames (such as red when combined with strontium nitrate), noise (with aluminum, in "flash and sound" mixtures), and light (in photoflash mixtures with magnesium).

Ammonium Perchlorate (NH_4ClO_4)

The "newest" oxidizer to appear in pyrotechnics, ammonium perchlorate has found considerable use in modern solid-fuel rocket propellants and in the fireworks industry. The space shuttle alone uses approximately two million pounds of solid fuel per launch; the mixture is 70% ammonium perchlorate, 16% aluminum metal, and 14% organic polymer.

Ammonium perchlorate undergoes a complex chemical reaction on heating, with decomposition occurring over a wide range, beginning near 200°C. Decomposition occurs prior to melting, so a liquid state is not produced – the solid starting material goes directly to gaseous decomposition products. The decomposition reaction is reported by Shimizu [2] to be

$$2\ NH_4ClO_4 \xrightarrow{heat} N_2 + 3\ H_2O + 2\ HCl + 2.5\ O_2$$

This equation corresponds to the evolution of 80 grams (2.5 moles) of oxygen gas per 2 moles (235 grams) of NH_4ClO_4, giving an "active oxygen" content of 34% (versus 39.2% for $KClO_3$ and 46.2% for $KClO_4$). The decomposition reaction, above 350°C, is reported to be considerably more complex [9].

$$10\ NH_4ClO_4 \xrightarrow{heat} 2.5\ Cl_2 + 2\ N_2O + 2.5\ NOCl + HClO_4 + 1.5\ HCl + 18.75\ H_2O + 1.75\ N_2 + 6.38\ O_2$$

Mixtures of ammonium perchlorate with fuels can produce high temperatures when ignited, and the hydrogen chloride (HCl) liberated during the reaction can aid in the production of colors. These two factors make ammonium perchlorate a good oxidizer for colored flame compositions (see Chapter 7).

Ammonium perchlorate is more hygroscopic than potassium nitrate or potassium chlorate, and some precautions should be taken to keep mixtures dry. The hygroscopicity problem can be substantial if a given composition also contains potassium nitrate, or even comes in contact with a potassium nitrate-containing mixture. Here, the reaction

$$NH_4ClO_4 + KNO_3 \xrightarrow{H_2O} KClO_4 + NH_4NO_3$$

can occur, especially in the presence of moisture. The exchange product, ammonium nitrate (NH_4NO_3) is *very* hygroscopic, and ignition problems may well develop [2]. Also, ammonium perchlorate should not be used in combination with a chlorate-containing compound, due to the possible formation of unstable ammonium chlorate in the presence of moisture.

Magnesium metal should also be avoided in ammonium perchlorate compositions. Here, the reaction

$$2\ NH_4ClO_4 + Mg \rightarrow 2\ NH_3 + Mg(ClO_4)_2 + H_2 + \text{heat}$$

can occur in the presence of moisture. Spontaneous ignition may occur if the heat buildup is substantial.

Under severe initiation conditions, ammonium perchlorate can be made to explode by itself [10]. Mixtures of ammonium perchlorate with sulfur and antimony sulfide are reported to be considerably more shock sensitive than comparable $KClO_3$ compositions [2]. Ammonium perchlorate can be used to produce excellent colors, with little solid residue, but care must be exercised at all times with this oxidizer. The explosive properties of this material suggest that minimum amounts of bulk composition should be prepared at one time, and large quantities should not be stored at manufacturing sites.

Strontium Nitrate [$Sr(NO_3)_2$]

This material is rarely used as the *only* oxidizer in a composition, but is commonly combined with potassium perchlorate in red flame mixtures. It is a white crystalline solid with a melting point of approximately 570°C. It is somewhat hygroscopic, so moisture should be avoided when using this material.

Near its melting point, strontium nitrate decomposes according to

$$Sr(NO_3)_2 \rightarrow SrO + NO + NO_2 + O_2$$

Strontium nitrite – $Sr(NO_2)_2$ – is formed as an intermediate in this decomposition reaction, and a substantial quantity of the nitrite can be found in the ash of low flame temperature mixtures [2]. At higher reaction temperatures, the decomposition is

$$Sr(NO_3)_2 \rightarrow SrO + N_2 + 2.5\ O_2$$

This is a strongly endothermic reaction, with a heat of reaction of +92 kcal, and corresponds to an active oxygen content of 37.7%. Little ash is produced by this high-temperature process, which occurs in mixtures containing magnesium or other "hot" fuels.

Barium Nitrate [$Ba(NO_3)_2$]

Barium nitrate is a white, crystalline, non-hygroscopic material with a melting point of approximately 592°C. It is commonly used as the principal oxidizer in green flame compositions, gold sparklers, and in photoflash mixtures in combination with potassium perchlorate.

At high reaction temperatures, barium nitrate decomposes according to

$$Ba(NO_3)_2 \rightarrow BaO + N_2 + 2.5\ O_2$$

This reaction corresponds to 30.6% available oxygen. At lower reaction temperatures, barium nitrate produces nitrogen oxides (NO and NO_2) instead of nitrogen gas, as does strontium nitrate [2].

Mixtures containing barium nitrate as the sole oxidizer are typically characterized by high ignition temperatures, relative to potassium nitrate and potassium chlorate compositions. The higher melting point of barium nitrate is responsible for these higher ignition values.

Other Oxidizers

A variety of other oxidizers are also occasionally used in high-energy mixtures, generally with a specific purpose in mind. Barium chlorate – $Ba(ClO_3)_2$ – for example is used in some green flame compositions. These mixtures can be very sensitive, however, and great care must be used during mixing, loading, and storing. Barium chlorate can be used to produce a beautiful green flame, though.

Barium chlorate is interesting because it exists as a *hydrate* when crystallized from a water solution. It has the formula $Ba(ClO_3)_2 \cdot H_2O$. Water molecules are found in the crystalline lattice in a one-to-one ratio with barium ions. The molecular weight of the hydrate is 322.3 ($Ba + 2\ ClO_3 + H_2O$), so the water must be included in stoichiometry calculations. On heating, the water is driven off at 120°C, producing anhydrous $Ba(ClO_3)_2$, which later melts at 414°C. The thermal decomposition of barium chlorate is strongly *exothermic* (-28 kcal/mole). This value, considerably greater than that of potassium chlorate, causes barium chlorate mixtures to be very sensitive to friction, heat, and other ignition stimuli.

Iron oxide (hematite, Fe_2O_3) is used in certain mixtures where a high ignition temperature and a substantial quantity of molten slag (and lack of gaseous product) are desired. The *thermite* reaction,

$$Fe_2O_3 + 2\ Al \rightarrow Al_2O_3 + 2\ Fe$$

is an example of this type of reaction, and can be used to do pyrotechnic welding. The melting point of Fe_2O_3 is 1565°C, and the ignition temperature of thermite mix is above 800°C. A reaction temperature of approximately 2400°C is reached, and 950 calories of heat is evolved per gram of composition [2, 5].

Other oxidizers, including barium chromate ($BaCrO_4$), lead chromate ($PbCrO_4$), sodium nitrate ($NaNO_3$), lead dioxide (PbO_2), and barium peroxide (BaO_2) will also be encountered in subsequent chapters. Bear in mind that reactivity and ease of ignition are often related to the melting point of the oxidizer, and the volatility of the reaction products determins the amount of gas that will be formed from a given oxidizer/fuel combination. Table 3.2 contains the physical and chemical properties of the common oxidizers, and Table 5.8 lists the melting and boiling points of some of the common reaction products.

Shidlovskiy has pointed out that metal-fluorine compounds should also have good oxidizer capability. For example, the reaction

$$FeF_3 + Al \rightarrow AlF_3 + Fe$$

is quite exothermic ($\Delta H = -70$ kcal). However, the lack of stable, economical metal fluorides of the proper reactivity has limited research in this direction [1].

FUELS

Requirements

In addition to an oxidizer, pyrotechnic mixtures will also contain a good fuel – or electron donor – that reacts with the liberated oxygen to produce an oxidized product plus heat. This heat will enable the high-energy chemist to produce any of a variety of possible effects – color, motion, light, smoke, or noise.

The desired pyrotechnic effect must be carefully considered when a fuel is selected to pair with an oxidizer for a high-energy mixture. Both the flame temperature that will be produced and the nature of the reaction products are important factors. The requirements for some of the major pyrotechnic categories are:

1. *Propellants*: A combination producing high temperature, a large volume of low molecular weight gas, and a rapid burnning rate is needed. Charcoal and organic compounds are often found in these compositions because of the gaseous products formed upon their combustion.
2. *Illuminating compositions*: A high reaction temperature is mandatory to achieve intense light emission, as is the presence in the flame of strong light-emitting species. Magnesium is commonly found in such mixtures due to its good heat output. The production of incandescent magnesium oxide particles in the flame aids in achieving good light intensity. Atomic sodium, present in vapor form in a flame, is a very strong light emitter, and sodium emission dominates the light output from the widely used sodium nitrate/magnesium compositions.
3. *Colored flame compositions*: A high reaction temperature produces maximum light intensity, but color *quality* depends upon having the *proper* emitters present in the flame, with a minimum of solid and liquid particles present that are emitting a broad spectrum of "white" light. Magnesium is sometimes added to colored flame mixtures to obtain higher intensity, but the color quality may suffer due to broad emission from MgO particles. Organic fuels (red gum, dextrine, etc.) are found in most color mixtures used in the fireworks industry.
4. *Colored smoke compositions*: Gas evolution is needed to disperse the smoke particles. High temperatures are *not* desirable here because decomposition of the organic dye molecules will occur. Metals are not found in these mixtures. Low heat fuels such as sulfur and sugars are commonly employed.
5. *Ignition compositions*: Hot solid or liquid particles are desirable in igniter and first-fire compositions to insure the transfer of sufficient heat to ignite the main composition. Fuels producing mainly gaseous products are not commonly used.

A good fuel will react with oxygen (or a halogen like fluorine or chlorine) to form a stable compound, and substantial heat will be evolved. The considerable strength of the metal-oxygen and metal-halogen bonds in the reaction products accounts for the excellent fuel properties of many of the metallic elements.

A variety of materials can be used, and the choice of material will depend on a variety of factors – the amount of heat output required, rate of heat release needed, cost of the materials, stability of the fuel and fuel/oxidizer pair, and amount of gaseous product desired. Fuels can be divided into three main categories: metals, non-metallic elements, and organic compounds.

Metals

A good metallic fuel resists air oxidation and moisture, has a high heat output per gram, and is obtainable at moderate cost in fine particle sizes. Aluminum and magnesium are the most widely used materials. Titanium, zirconium, and tungsten are also used, especially in military applications.

The alkali and alkaline earth metals – such as sodium, potassium, barium, and calcium -- would make excellent high-energy fuels, but, except for magnesium, they are *too* reactive with moisture and atmospheric oxygen. Sodium metal, for example, reacts violently with water and must be stored in an inert organic liquid, such as xylene, to minimize decomposition.

A metal can initially be screened for pyrotechnic possibilities by an examination of its standard reduction potential (Table 2.5). A readily oxidizable material will have a large, negative value, meaning it possesses little tendency to *gain* electrons and a significant tendency to *lose* them. Good metallic fuels will also be reasonably lightweight, producing high calories/gram values when oxidized. Table 3.4 lists some of the common metallic fuels and their properties.

Aluminum (Al)

The most widely used metallic fuel is probably aluminum, with magnesium running a close second. Aluminum is reasonable in cost, lightweight, stable in storage, available in a variety of particle shapes and sizes, and can be used to achieve a variety of effects.

Aluminum has a melting point of 660°C and a boiling point of approximately 2500°C. Its heat of combustion is 7.4 kcal/gram.

TABLE 3.4 Properties of Metallic Fuels

Element	Symbol	Atomic weight	Melting point, °C[a]	Boiling point, °C[a]	Heat of combustion, kcal/gram[b]	Combustion product	Grams of fuel consumed per gram of O
Aluminum	Al	27.0	660	2467	7.4	Al_2O_3	1.12
Iron	Fe	55.8	1535	2750	1.8	Fe_2O_3	2.32
Magnesium	Mg	24.3	649	1107	5.9	MgO	1.52
"Magnalium"	Mg/Al (usually a 50/50 alloy)	-	460	-	-	MgO/Al_2O_3	1.32
Titanium	Ti	47.9	1660	3287	4.7	TiO_2	1.50
Tungsten	W	183.8	3410	5660	1.1	WO_3	3.83
Zinc	Zn	65.4	420	907	1.3	ZnO	4.09
Zirconium	Zr	91.2	1852	4377	2.9	ZrO_2	2.85

[a]Reference 4.
[b]Reference 1.

Aluminum is available in either "flake" or "atomized" form. The "atomized" variety consists of spheroidal particles. Spheres yield the minimum surface area (and hence minimum reactivity) for a given particle size, but this form will be the most *reproducible* in performance from batch to batch. Atomized aluminum, rather than the more reactive flake material, is used by the military for heat and light-producing compositions because the variation in performance from shipment to shipment is usually less.

Large flakes, called "flitter" aluminum, are widely used by the fireworks industry to produce bright white sparks. A special "pyro" grade of aluminum is also available from some suppliers. This is a dark gray powder consisting of small particle sizes and high surface area and it is *extremely* reactive. It is used to produce explosive mixtures for fireworks, and combinations of oxidizers with this "pyro" aluminum should only be prepared by skilled personnel, and only made in small batches. Their explosive power can be substantial, and they can be quite sensitive to ignition.

Aluminum surfaces are readily oxidized by the oxygen in air, and a tight surface coating of aluminum oxide (Al_2O_3) is formed that protects the inner metal from further oxidation. Hence, aluminum powder can be stored for extended periods with little loss of reactivity due to air oxidation. Metals that form a loose oxide coating on exposure to air – iron, for example – are not provided this surface protection, and extensive decomposition can occur during storage unless appropriate precautions are taken.

Compositions made with aluminum tend to be quite stable. However, moisture must be excluded if the mixture also contains a nitrate oxidizer. Otherwise, a reaction of the type

$$3\,KNO_3 + 8\,Al + 12\,H_2O \rightarrow 3\,KAlO_2 + 5\,Al(OH)_3 + 3\,NH_3$$

can occur, evolving heat and ammonia gas. This reaction is accelerated by the alkaline medium generated as the reaction proceeds, and autoignition is possible in a confined situation. A small quantity of a weak acid such as boric acid (H_3BO_3) can effectively retard this decomposition by neutralizing the alkaline products and maintaining a weakly acidic environment. The hygroscopicity of the oxidizer is also important in this decomposition process. Sodium nitrate and aluminum can not be used together, due to the high moisture affinity of $NaNO_3$, unless the aluminum powder is coated with a protective layer of wax or similar material. Alternatively, the product can be sealed in a moisture-proof packaging to exclude any water [1]. Potassium nitrate/aluminum compositions must be kept quite dry in storage to avoid decomposition problems, but mixtures of aluminum and non-hygroscopic barium nitrate can be stored with a minimum of precautions, as long as the composition does not actually get wet. Mixtures of magnesium metal with nitrate salts do not have this alkaline-catalyzed decomposition problem. A magnesium hydroxide [$Mg(OH)_2$] coating on the metal surface apparently protects it from further reaction. This protection is not provided to aluminum metal by the alkaline-soluble aluminum hydroxide, $Al(OH)_3$.

Magnesium (Mg)

Magnesium is a very reactive metal and makes an excellent fuel under the proper conditions. It is oxidized by moist air to form magnesium hydroxide, $Mg(OH)_2$, and it readily reacts with all acids, including weak species such as vinegar (5% acetic acid) and boric acid. The reactions of magnesium with water and an acid (HX) are shown below:

Water: $Mg + 2\ H_2O \rightarrow Mg(OH)_2 + H_2$

Acids (HX): $Mg + 2\ HX \rightarrow MgX_2 + H_2$ (X = Cl, NO_3, etc.)

Even the ammonium ion, NH_4^+, is acidic enough to react with magnesium metal. Therefore, ammonium perchlorate and other ammonium salts should not be used with magnesium unless the metal surface is coated with linseed oil, paraffin, or a similar material.

Chlorate and perchlorate salts, in the presence of moisture, will oxidize magnesium metal, destroying any pyrotechnic effect during storage. Nitrate salts appear to be considerably more stable with magnesium [2]. Again, coating the metal with an organic material – such as paraffin – will increase the storage lifetime of the composition. A coating of potassium dichromate on the surface of the magnesium has also been recommended to aid in stability [2], but the toxicity of this material makes it of questionable value for industrial applications.

Magnesium has a heat of combustion of 5.9 kcal/gram, a melting point of 649°C, and a low boiling point of 1107°C. This low boiling point allows excess magnesium in a mixture to vaporize and burn with oxygen in the air, providing additional heat (and light) in flare compositions. No heat absorption is required to decompose an oxidizer when this excess magnesium reacts with atmospheric oxygen; hence, the extra heat gained by incorporating the excess magnesium into the mixture is substantial.

Magnesium metal is also capable of reacting with other metal ions in an electron-transfer reaction, such as

$$Cu^{+2} + Mg \rightarrow Cu + Mg^{+2}$$

This process becomes much more probable if a composition is moistened, again pointing out the variety of problems that can be created if water is added to a magnesium-containing mixture. The standard potential for the Cu^{+2}/Mg system is +2.72 volts, indicating a very spontaneous process. Therefore, Cu^{+2}, Pb^{+2}, and other readily-reducible metal ions must not be used in magnesium-containing compositions.

"Magnalium" (Magnesium-Aluminum Alloy)

A material finding increasing popularity in pyrotechnics is the 50/50 alloy of magnesium and aluminum, termed "magnalium." Shimizu reports that this material is a solid solution of Al_3Mg_2 in Al_2Mg_3, with a melting point of 460°C [2]. The alloy is considerably more stable than aluminum metal when combined with nitrate salts, and reacts much more slowly than magnesium metal with weak acids. It therefore offers stability advantages over both of its component materials.

The Chinese make wide use of magnalium in fireworks items to produce attractive white sparks and "crackling" effects. Shimizu also reports that a branching spark effect can be produced using magnalium with a black powder-type composition [2].

Iron

Iron, in the form of fine filings, will burn and can be used to produce attractive gold sparks, such as in the traditional wire sparkler. The small percentage (less than 1%) of carbon in steel can cause an attractive branching of the sparks due to carbon dioxide gas formation as the metal particles burn in air.

Iron filings are quite unstable on storage, however. They readily convert to iron oxide (rust – Fe_2O_3) in moist air, and filings are usually coated with a paraffin-type material prior to use in a pyrotechnic mixture.

Other Metals

Titanium metal (Ti) offers some attractive properties to the high-energy chemist. It is quite stable in the presence of moisture and most chemicals, and produces brilliant silver-white spark and light effects with oxidizers. Lancaster feels that it is a safer material to use than either magnesium or aluminum, and recommends that it be used in place of iron filings in fireworks "fountain" items, due to its greater stability [11]. Cost and lack of publicity seem to be the major factors keeping titanium from being a much more widely used fuel.

Zirconium (Zr) is another reactive metal, but its considerable expense is a major problem restricting its wider use in high-energy compositions. It is easily ignited – and therefore quite hazardous – as a fine powder, and must be used with great care.

Non-Metallic Elements

Several readily-oxidized nonmetallic elements have found widespread use in the field of pyrotechnics. The requirements again are stability to air and moisture, good heat-per-gram output, and reasonable cost. Materials in common use include sulfur, boron, silicon, and phosphorus. Their properties are summarized in Table 3.5.

Sulfur

The use of sulfur as a fuel in pyrotechnic compositions dates back over one thousand years, and the material remains a widely-used component in black powder, colored smoke mixtures, and fireworks compositions. For pyrotechnic purposes, the material termed "flour of sulfur" that has been crystallized from molten sulfur is preferred. Sulfur purified by sublimation – termed "flowers of sulfur" – often contains significant amounts of oxidized, acidic impurities and can be quite hazardous in high-energy mixtures, especially those containing a chlorate oxidizer [11].

Sulfur has a particularly low (119°C) melting point. It is a rather poor fuel in terms of heat output, but it frequently plays another very important role in pyrotechnic compositions. It can function as a "tinder," or fire starter. Sulfur undergoes exothermic reactions at low temperature with a variety of oxidizers, and this heat output can be used to trigger other, higher-energy reactions with better fuels. Sulfur's low melting point provides a liquid phase, at low temperature, to assist the ignition process. The presence of sulfur, even in small percentage, can dramatically affect the ignitibility and ignition temperature of high-energy mixtures. Sulfur, upon combustion, is converted to sulfur dioxide gas and to sulfate salts (such as potassium sulfate – K_2SO_4). Sulfur is also found to act as an oxidizer in some

TABLE 3.5 Properties of the Non-Metallic Elements Used as Fuels in High-Energy Mixtures

Element	Symbol	Atomic weight	Melting point, °C[a]	Boiling point, °C[a]	Heat of combustion (kcal/g)[b,c]	Combustion product	Grams of fuel consumed per gram of O
Boron	B	10.8	2300	2550	14.0	B_2O_3	.45
Carbon (charcoal)	C	12 (approx.)	Dec.	-	7.8	CO_2	.38
Phosphorus (red)	P	31.0	590	Sublimes	5.9	P_2O_5	.78
Phosphorus (yellow)	P_4	124.0 (as P_4)	44	-	5.9	P_2O_5	.78
Silicon	Si	28.1	1410	2355	7.4	SiO_2	.88
Sulfur	S	32.1	119	445	2.2	SO_2	1.00

[a]Reference 4.
[b]Reference 1.
[c]Reference 2.

mixtures, winding up as the sulfide ion (S^{-2}) in species such as potassium sulfide (K_2S), a detectable component of black powder combustion residue.

When present in large excess, sulfur may volatilize out of the burning mixture as yellowish-white smoke. A 1:1 ratio of potassium nitrate and sulfur makes a respectable smoke composition employing this behavior.

Boron

Boron is a stable element, and can be oxidized to yield good heat output. The low atomic weight of boron (10.8) makes it an excellent fuel on a calories/gram basis. Boron has a high melting point (2300°C), and it can prove hard to ignite when combined with a high-melting oxidizer. With low-melting oxidizers, such as potassium nitrate, boron ignites more readily yielding good heat production. The low melting point of the oxide product (B_2O_3) can interfere with the attainment of high reaction temperatures, however, [1].

Boron is a relatively expensive fuel, but it frequently proves acceptable for use on a cost basis because only a small percentage is required (remember, it has a low atomic weight). For example, the reaction

$$BaCrO_4 + B \rightarrow \text{products } (B_2O_3,\ BaO,\ Cr_2O_3)$$

burns well with only 5% by weight boron in the composition [5, 6]. Boron is virtually unknown in the fireworks industry, but is a widely-used fuel in igniter and delay compositions for military and aerospace applications.

Silicon

In many ways similar to boron, silicon is a safe, relatively inexpensive fuel used in igniter and delay compositions. It has a high melting point (1410°C), and combinations of this material with a high-melting oxidizer may be difficult to ignite. The oxidation product, silicon dioxide (SiO_2), is high melting and, importantly, is environmentally acceptable.

Phosphorus

Phosphorus is an example of a material that is *too* reactive to be of any general use as a pyrotechnic fuel, although it is increasingly being employed in military white smoke compositions, and it has traditionally been used in toy pistol caps and trick noisemakers ("party poppers").

Phosphorus is available in two forms, white (or yellow) and red. White phosphorus appears to be molecular, with a formula of P_4. It is a waxy solid with a melting point of 44°C, and ignites spontaneously on exposure to air. It must be kept cool and is usually stored under water. It is highly toxic in both the solid and vapor form and causes burns on contact with the skin. Its use in pyrotechnics is limited to incendiary and white smoke compositions. The white smoke consists of the combustion product, primarily phosphoric acid (H_3PO_4).

Red phosphorus is somewhat more stable, and is a reddish-brown powder with a melting point of approximately 590°C (in the absence of air). In the presence of air, red phosphorus ignites near 260°C [2]. Red phosphorus is insoluble in water. It is easily ignited by spark or friction, and is quite hazardous any time it is mixed with oxidizers or flammable materials. Its fumes are highly toxic [3].

Red phosphorus is mixed as a water slurry with potassium chlorate for use in toy caps and noisemakers. These mixtures are quite sensitive to friction, impact, and heat, and a large amount of such mixtures must never be allowed to dry out in bulk form. Red phosphorus is also used in white smoke mixtures, and several examples can be found in Chapter 8.

Sulfide Compounds

Several metallic sulfide compounds have been used as fuels in pyrotechnic compositions. Antimony trisulfide, Sb_2S_3, is a reasonably low-melting material (m.p. 548°C) with a heat of combustion of approximately 1 kcal/gram. It is easily ignited and can be used to aid in the ignition of more difficult fuels, serving as a "tinder" in the same way that elemental sulfur does. It has been used in the fireworks industry for white fire compositions and has been used in place of sulfur in "flash and sound" mixtures with potassium perchlorate and aluminum.

Realgar (arsenic disulfide, As_2S_2) is an orange powder with a melting point of 308°C and a boiling point of 565°C [2]. Due to its low boiling point, it has been used in yellow smoke compositions (in spite of its toxicity!), and has also been used to aid in the ignition of difficult mixtures.

The use of all arsenic compounds -- including realgar -- is prohibited in "common fireworks" (the type purchased by individuals) by regulations of the U.S. Consumer Product Safety Commission [12].

Organic Fuels

A variety of organic (carbon-containing) fuels are commonly employed in high-energy compositions. In addition to providing heat, these materials also generate significant gas pressure through the production of carbon dioxide (CO_2) and water vapor in the reaction zone.

The carbon atoms in these molecules are oxidized to carbon dioxide if sufficient oxygen is present. Carbon monoxide (CO) or elemental carbon are produced in an oxygen-deficient atmosphere, and a "sooty" flame is observed if a substantial amount of carbon is generated. The hydrogen present in organic compounds winds up as water molecules. For a fuel of formula $C_xH_yO_z$, x moles of CO_2 and y/2 moles of water will be produced per mole of fuel that is burned. To completely combust this fuel, x + y/2 moles of oxygen gas (2x + y moles of oxygen atoms) will be required. The amount of oxygen that must be provided by the oxidizer in a high-energy mixture is reduced by the presence of oxygen atoms in the fuel molecule. The balanced equation for the combustion of glucose is shown below:

$$C_6H_{12}O_6 + 6\ O_2 \rightarrow 6\ CO_2 + 6\ H_2O$$

Only six oxygen molecules are required to oxidize one glucose molecule, due to the presence of six "internal" oxygen atoms in glucose. There are 18 oxygen atoms on both sides of the balanced equation.

A fuel that contains only carbon and hydrogen – termed a hydrocarbon – will require more moles of oxygen for complete combustion than will an equal weight of glucose or other oxygen-containing compound. A greater weight of oxidizer is therefore required per gram of fuel when a hydrocarbon-type material is used.

The grams of oxygen needed to completely combust one gram of a given fuel can be calculated from the balanced chemical equation. Table 3.6 lists the oxygen requirement for a variety of organic fuels. A sample calculation is shown in Figure 3.1.

To determine the proper ratio of oxidizer to fuel for a stoichiometric composition, the grams of oxygen required by a given fuel (Tables 3.4-3.6) must be matched with the grams of oxygen delivered by the desired oxidizer (given in Table 3.2). For the reaction between potassium chlorate ($KClO_3$) and glucose ($C_6H_{12}O_6$), 2.55 grams of $KClO_3$ donates 1.00 grams of oxygen, and 0.938 grams of glucose consumes 1.00 grams of oxygen. The proper weight ratio of potassium chlorate to glucose is therefore 2.55: 0.938, and the stoichiometric mixture should be 73.1% $KClO_3$ and

TABLE 3.6 Properties of Some Common Organic Fuels

Compound	Formula	Molecular weight	Melting point, °C[a]	Grams of fuel consumed per gram of O	Heat of combustion
Monomers					kcal/mole[a]
Lactose	$C_{12}H_{22}O_{11} \cdot H_2O$	360.3	202	0.94	1351
Naphthalene	$C_{10}H_8$	128.2	80.5	0.33	1232[b]
Shellac	Primarily $C_{16}H_{32}O_5$	ca. 304	ca. 120	ca. 0.44	-
Stearic acid	$C_{18}H_{36}O_2$	284.5	69.5	0.34	2712
Sucrose	$C_{12}H_{22}O_{11}$	342.3	188 (decomposes)	0.89	1351
Polymers					cal/gram[a]
Dextrine	$(-C_6H_{10}O_5-)_n \cdot H_2O$	-	Decomposes	ca. 0.84	ca. 4179
Laminac	Polyester/styrene copolymer	-	ca. 200 (decomposes)	-	-
Nitrocellulose	$(C_6H_{10-x}O_{5-x}(ONO_2)_x)_n$	-	ca. 200 (decomposes)	-	2409
Polyvinyl chloride	$(-CH_2CHCl-)_n$	ca. 250,000	ca. 80 (softens)	ca. 0.78	4375
Starch	$(C_6H_{10}O_5)_n$	-	Decomposes	ca. 0.84	4179

[a] Reference 3.
[b] Reference 4.

Equation:	$C_6H_{12}O_6$ +	6 O_2 →	6 CO_2 +	6 H_2O
moles	1	6	6	6
grams	180	192	264	108
grams/gram O	0.938	1.00		

[Obtained by setting up the ratio

$$\frac{180}{192} = \frac{X}{1.00}$$

and solving

$$X = \frac{(180)(1.00)}{192} = 0.938]$$

FIG. 3.1 Calculation of oxygen demand. The quantity of oxygen consumed during the combustion of an organic fuel can be calculated by first balancing the equation for the overall reaction. Each carbon atom in the fuel converts to a carbon dioxide molecule (CO_2), and every two hydrogen atoms yield a water molecule. The oxygen required to burn the fuel is determined by adding up all of the atoms of oxygen in the products and then subtracting the oxygen atoms (if any) present in the fuel molecule. The difference is the number of oxygen atoms that must be supplied by the atmosphere (or by an oxidizer). This number is then divided by 2 to obtain the number of O_2 molecules needed. The coefficients can then be multiplied by the appropriate molecular weights to obtain the number of grams involved.

26.9% glucose by weight. An identical answer is obtained if the chemical equation for the reaction between $KClO_3$ and glucose is balanced and the molar ratio then converted to a weight ratio:

$$C_6H_{12}O_6 + 4\ KClO_3 \rightarrow 6\ CO_2 + 6\ H_2O + 4\ KCl$$

	$C_6H_{12}O_6$	$KClO_3$
Moles:	1	4
Grams:	180	490
Weight %:	26.9	73.1

The more highly oxidized — or oxygen rich — a fuel is, the smaller its heat output will be when combusted. The flame temperature will also be lower for compositions using the highly-oxidized fuel. Also, fuels that exist as hydrates (containing water of crystallization) will evolve less heat than similar, nonhydrated species due to the absorption of heat required to vaporize the water present in the hydrates.

Two "hot" organic fuels are shellac and red gum. Shellac, secreted by an Asian insect, contains a high percentage of trihydroxypalmitic acid – $CH_3(CH_2)_{11}(CHOH)_3COOH$ [2]. This molecule contains a low percentage of oxygen and produces a high heat/gram value. Red gum is a complex mixture obtained from an Australian tree, with excellent fuel characteristics and a low melting point to aid in ignition.

Charcoal is another organic fuel, and has been employed in high-energy mixtures for over a thousand years. It is prepared by heating wood in an air-free environment; volatile products are driven off and a residue that is primarily carbon remains. Shimizu reports that a highly-carbonized sample of charcoal showed a 91:3:6 ratio of C, H, and O atoms [2].

The pyrotechnic behavior of charcoal may vary greatly depending upon the type of wood used to prepare the material. The surface area and extent of conversion to carbon may vary widely from wood to wood and batch to batch, and each preparation must be checked for proper performance [13]. Historically, willow and alder have been the woods preferred for the preparation of charcoal by black powder manufacturers.

Charcoal is frequently the fuel of choice when high heat and gas output as well as a rapid burning rate are desired. The addition of a small percentage of charcoal to a sluggish composition will usually accelerate the burning rate and facilitate ignition.

Larger particles of charcoal in a pyrotechnic mixture will produce attractive orange sparks in the flame, a property that is often used to advantage by the fireworks industry.

Carbohydrates

The carbohydrate family consists of a large number of naturally-occurring oxygen-rich organic compounds. The simplest carbohydrates – or "sugars" – have molecular formulas fitting the pattern $(C \cdot H_2O)_n$, and appeared to early chemists to be "hydrated carbon." The more complex members of the family deviate from this pattern slightly.

Examples of common sugars include glucose ($C_6H_{12}O_6$), lactose ($C_{12}H_{22}O_{11}$), and sucrose ($C_{12}H_{22}O_{11}$). Starch is a complex polymer composed of glucose units linked together. The molecular formula of starch is similar to $(C_6H_{10}O_5)_n$, and the molecular weight of starch is typically greater than one million. Reaction with acid breaks starch down into smaller units. Dextrine, a widely-used pyrotechnic fuel and binder, is partially-hydrolyzed starch. Its molecular weight, solubility, and chemical behavior may vary considerably from supplier to supplier and from batch to batch. The testing of all new shipments of dextrine is required in pyrotechnic production.

The simpler sugars are used as fuels in various pyrotechnic mixtures. They tend to burn with a colorless flame and give off less heat per gram than less-oxidized organic fuels. Lactose is used with potassium chlorate in some colored smoke mixtures to produce a low-temperature reaction capable of volatilizing an organic dye with minimum decomposition of the complex dye molecule. The simpler sugars can be obtained in high purity at moderate cost, making them attractive fuel choices. Toxicity problems tend to be minimal with these fuels, also.

Other Organic Fuels

The number of possible organic fuels is enormous. Considerations in selecting a candidate are:

1. *Extent of oxidation*: This will be a primary factor in the heat output/gram of the fuel.
2. *Melting point*: A low melting point can aid in ignitibility and reactivity; too low a melting point can cause production and storage problems. 100°C might be a good minimum value.
3. *Boiling point*: If the fuel is quite volatile, the storage life of the mixture will be brief unless precautions are taken in packaging to prevent loss of the material.
4. *Chemical stability*: An ideal fuel should be available commercially in a high state of purity, and should maintain that high purity during storage. Materials that are easily air-oxidized, such as aldehydes, are poor fuel choices.
5. *Solubility*: Organic fuels frequently double as binders, and some solubility in water, acetone, or alcohol is required to obtain good binding behavior.

Materials that have been used in pyrotechnic mixtures include nitrocellulose, polyvinyl alcohol, stearic acid, hexamethylenetetramine, kerosene, epoxy resins, and unsaturated polyester resins such as Laminac. The properties of most of these fuels can be found in a handbook prepared by the U.S. Army [3]. Table 3.6 contains information on a variety of organic compounds that are of interest to the high-energy chemist.

BINDERS

A pyrotechnic composition will usually contain a small percentage of an organic polymer that functions as a binder, holding all of the components together in a homogeneous blend. These binders, being organic compounds, will also serve as fuels in the mixture.

Without the binder, materials might well segregate during manufacture and storage due to variations in density and particle size. The granulation process, in which the oxidizer, fuel, and other components are blended with the binder (and usually a suitable solvent) to produce grains of homogeneous composition, is a critical step in the manufacturing process. The solvent is evaporated following granulation, leaving a dry, homogeneous material.

Dextrine is widely used as a binder in the fireworks industry. Water is used as the wetting agent for dextrine, avoiding the cost associated with the use of organic solvents.

Other common binders include nitrocellulose (acetone as the solvent), polyvinyl alcohol (used with water), and Laminac (an unsaturated polyester crosslinked with styrene -- the material is a liquid until cured by catalyst, heat, or both, and no solvent is required). Epoxy binders can also be used in liquid form during the mixing process and then allowed to cure to leave a final, rigid product.

In selecting a binder, the chemist seeks a material that will provide good homogeneity with the use of a minimum of polymer. Organic materials will reduce the flame temperatures of compositions containing metallic fuels, and they can impart an orange color to flames if incomplete combustion of the binder occurs and carbon forms in the flame. A binder should be neutral and nonhygroscopic to avoid the problems that water and an acidic or basic environment can introduce. For example, magnesium-containing mixtures require the use of a non-aqueous binder/solvent system, because of the reactivity of magnesium metal towards water. When iron is used in a composition, pretreatment of the metal with wax or other protective coating is advisable, especially if an aqueous binding process is used.

RETARDANTS

Occasionally, a pyrotechnic mixture will function quite well and produce the desired effect, except for the fact that the burning rate is a bit too fast. A material is needed that will slow down the reaction without otherwise affecting performance. This can be accomplished by altering the ratio of ingredients (e.g., reducing the amount of fuel) or by adding an inert component to the composition. Excess metallic fuel is less effective as a "coolant" because of the ability of many fuels – such as magnesium – to react with the oxygen in air and liberate heat. Also, metals tend to be excellent heat conductors, and an increase in the metal percentage can speed up a reaction by facilitating heat transfer through the composition during the burning process.

Materials that decompose at elevated temperatures with the absorption of heat (endothermic decomposition) can work well as rate retardants. Calcium and magnesium carbonate, and sodium bicarbonate, are sometimes added to a mixture for this purpose.

$$CaCO_3 \text{ (solid)} \xrightarrow{\text{heat}} CaO \text{ (solid)} + CO_2 \text{ (gas)}$$

$$2\ NaHCO_3 \text{ (solid)} \rightarrow Na_2O \text{ (solid)} + H_2O \text{ (gas)} + 2\ CO_2 \text{ (gas)}$$

However, gas generation occurs that may or may not affect the performance of the mixture.

Although endothermic, these reactions are thermodynamically spontaneous at high temperature due to the favorable entropy change associated with the formation of random gaseous products from solid starting materials.

Inert diluents such as clay and diatomaceous earth can also be used to retard burning rates. These materials absorb heat and separate the reactive components, thereby slowing the pyrotechnic reaction.

REFERENCES

1. A. A. Shidlovskiy, *Principles of Pyrotechnics*, 3rd Ed., Moscow, 1964. (Translated by Foreign Technology Division, Wright-Patterson Air Force Base, Ohio, 1974.)
2. T. Shimizu, *Fireworks – The Art, Science & Technique*, pub. by T. Shimizu, distrib. by Maruzen Co., Ltd., Tokyo, 1981.

3. U.S. Army Material Command, Engineering Design Handbook, Military Pyrotechnic Series, Part Three, "Properties of Materials Used in Pyrotechnic Compositions," Washington, D.C., 1963 (AMC Pamphlet 706-187).
4. R. C. Weast (Ed.), *CRC Handbook of Chemistry and Physics*, 63rd Ed., CRC Press, Inc., Boca Raton, Fla., 1982.
5. H. Ellern, *Military and Civilian Pyrotechnics*, Chemical Publ. Co., Inc., New York, 1968.
6. T. J. Barton, et al., "Factors Affecting the Ignition Temperature of Pyrotechnics," *Proceedings, Eighth International Pyrotechnics Seminar*, IIT Research Institute, Steamboat Springs, Colorado, July, 1982, p. 99.
7. J. H. McLain, *Pyrotechnics from the Viewpoint of Solid State Chemistry*, The Franklin Institute Press, Philadelphia, Penna., 1980.
8. U.S. Department of Transportation, "Hazardous Materials Regulations," Code of Federal Regulations, Title 49, Part 173.
9. U.S. Army Material Command, Engineering Design Handbook, Military Pyrotechnic Series, Part One, "Theory and Application," Washington, D.C., 1967 (AMC Pamphlet 706-185).
10. D. Price, A. R. Clairmont, and I. Jaffee, "The Explosive Behavior of Ammonium Perchlorate," *Combustion and Flame*, *11*, 415 (1967).
11. R. Lancaster, *Fireworks Principles and Practice*, Chemical Publ. Co., Inc., New York, 1972.
12. U.S. Consumer Product Safety Commission, "Fireworks Devices," Code of Federal Regulations, Title 16, Part 1507.
13. J. E. Rose, "The Role of Charcoal in the Combustion of Black Powder," *Proceedings, Seventh International Pyrotechnics Seminar*, IIT Research Institute, Vail, Colorado, July, 1980, p. 543.

A pyrotechnician cautiously mixes a composition through a sieve to achieve homogeneity. Eye and respiratory protection are worn, and great care is taken throughout this critical phase of the manufacturing process. Sensitive compositions, as well as large quantities of any pyrotechnic mixture, should be blended remotely. (Fireworks by Grucci)

4
PYROTECHNIC PRINCIPLES

INTRODUCTION

The "secret" to maximizing the rate of reaction for a given pyrotechnic or explosive composition can be revealed in a single word – homogeneity. Any operation that increases the degree of intimacy of a high-energy mixture should lead to an enhancement of reactivity. Reactivity, in general, refers to the rate – in grams or moles per second – at which starting materials are converted into products.

The importance of intimate mixing was recognized as early as 1831 by Samuel Guthrie, Jr., a manufacturer of "fulminating powder" used to prime firearms. Guthrie's mixture was a blend of potassium nitrate, potassium carbonate, and sulfur, and he discovered that the performance could be dramatically improved if he first melted together the nitrate and carbonate salts, and then blended in the sulfur. He wrote, "By the previously melting together of the nitro and carbonate of potash, a more intimate union of these substances was effected than could possibly be made by mechanical means" [1]. However, he also experienced the hazards associated with maximizing reactivity, reporting, "I doubt whether, in the whole circle of experimental philosophy, many cases can be found involving dangers more appalling, or more difficult to be overcome, than melting fulminating powder and saving the product, and reducing the process to a business operation. I have had with it some eight or ten tremendous explosions, and in one of them I received, full in my face and eyes, the flame of a quarter of a pound of the composition, just as it had become thoroughly melted" [1]. An enormous debt is owed to these pioneers in high-energy chemistry who were willing to experiment in spite of the obvious hazards, and reported their results so others could build on their knowledge.

Varying degrees of homogeneity can be achieved by altering either the extent of mixing or the particle size of the various components. Striking differences in reactivity can result from changes in either of these, as Mr. Guthrie observed with his "fulminating powder."

A number of parameters related to burning behavior can be experimentally measured and used to report the "reactivity" or performance of a particular high-energy mixture [2]:

1. *Heat of reaction*: This value is expressed in units of calories (or kilocalories) per mole or calories per gram, and is determined using an instrument called a "calorimeter." One calorie of heat is required to raise the temperature of one gram of water by one degree (Celsius), so the temperature rise of a measured quantity of water, brought about by the release of heat from a measured amount of high-energy composition, can be converted into calories of heat. Depending upon the intended application, a mixture liberating a high, medium, or low value may be desired. Some representative heats of reaction are given in Table 4.1.
2. *Burning rate*: This is measured in units of inches, centimeters or grams per second for slow mixtures, such as delay compositions, and in meters per second for "fast" materials. Burning rates can be varied by altering the materials used, as well as the ratios of ingredients, as shown in Table 4.2. *Note*: Burning "rates" are also sometimes reported in units of seconds/cm or seconds/gram – the *inverse* of the previously-stated units. Always carefully read the units when examining burning rate data!
3. *Light intensity*: This is measured in candela or candlepower. The intensity is determined to a large extent by the temperature reached by the burning composition. Intensity will increase exponentially as the flame temperature rises, provided that no decomposition of the emitting species occurs.
4. *Color quality*: This will be determined by the relative intensities of the various wavelengths of light emitted by

TABLE 4.1 Representative Heats of Reaction for Pyrotechnic Systems[a]

Composition (% by weight)		$\Delta H_{reaction}$, kcal/gram	Application
Magnesium	50	2.0	Illuminating flare
Sodium nitrate, $NaNO_3$	44		
Laminac binder	6		
Potassium perchlorate, $KClO_4$	60	1.8	Photoflash
Aluminum	40		
Boron	25	1.6	Igniter
Potassium nitrate, KNO_3	75		
VAAR binder	1		
Potassium nitrate, KNO_3	71	1.0	Starter mixture
Charcoal	29		
Black powder	91	0.85	Flash and report
Aluminum	9		Military simulator
Barium chromate, $BaCrO_4$	85	0.5	Delay mixture
Boron	15		
Silicon	25	0.28	First fire mixture
Red lead oxide, Pb_3O_4	50		
Titanium	25		
Tungsten	50	0.23	Delay mixture
Barium chromate, $BaCrO_4$	40		
Potassium perchlorate, $KClO_4$	10		

[a]*Source*: F. L. McIntyre, "A Compilation of Hazard and Test Data for Pyrotechnic Compositions," Report AD-E400-496, U.S. Army Armament Research and Development Command, Dover, New Jersey, October 1980.

TABLE 4.2 Burning Rates of Binary Mixtures of Nitrate Oxidizers with Magnesium Metal[a]

% Oxidizer (by weight)	% Magnesium	Burning rate (inches/minute)[b] Barium nitrate oxidizer, $Ba(NO_3)_2$	Burning rate (inches/minute)[b] Potassium nitrate oxidizer, KNO_3
80	20	2.9	2.3
70	30	-	4.7
68	32	5.1	-
60	40	10.7	-
58	42	-	8.5
50	50	16.8	13.3
40	60	38.1	21.8
30	70	40.3	29.3
20	80	"Erratic"	26.4

[a]Reference 2.
[b]Loading pressure was 10,000 psi into 1.4 in^2 cases.

species present in the pyrotechnic flame. Only those wavelengths falling in the "visible" region of the electromagnetic spectrum will contribute to the color. An *emission spectrum*, showing the intensity of light emitted at each wavelength, can be obtained if the proper instrumentation – an emission spectrometer – is available (Figure 4.1).

5. *Volume of gas produced*: Gaseous products are frequently desirable when a high-energy mixture is ignited. Gas can be used to eject sparks, disperse smoke particles, and provide propellant behavior; when confined, gas can be used to create an explosion. Water, carbon monoxide and dioxide, and nitrogen are the main gases evolved from high-energy mixtures. The presence of organic compounds can

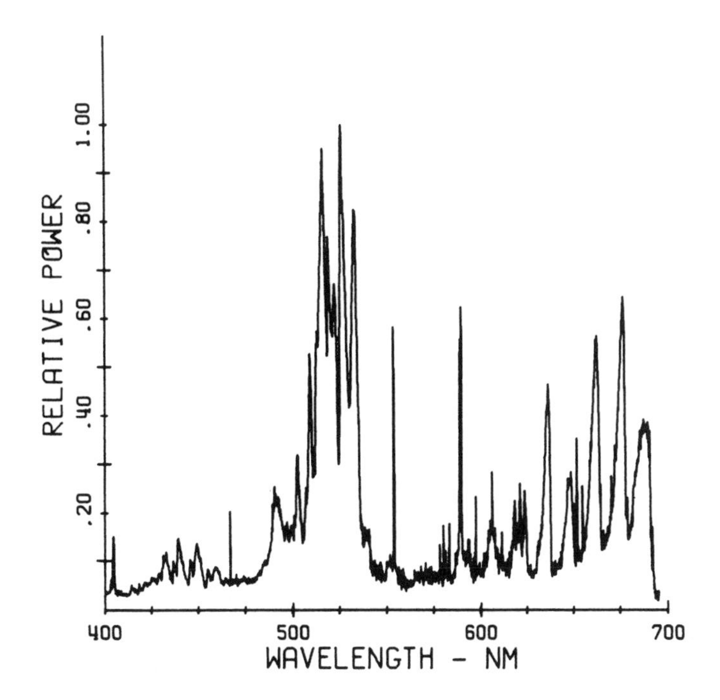

FIG. 4.1 Light output from a green flare. The radiant output from a burning pyrotechnic composition can be analyzed using an instrument known as a spectrophotometer. Energy output can be monitored as a function of wavelength. A good "white light" mixture will emit reasonably intense light over the entire visible region. Color will be produced when the emission is concentrated in a narrow portion of the visible range. The output from this flare falls largely between 500-540 nm -- the "green" portion of the visible spectrum. Green light emission is usually associated with the presence of a barium compound in the mixture, with molecular BaCl in the vapor state, typically the primary emitter of green light. The mixture producing this emission pattern consisted of potassium perchlorate (32.5%), barium nitrate (22.5%), magnesium (21%), copper powder (7%), polyvinyl chloride (12%), and 5% binder. *Source*: H. A. Webster III, "Visible Spectra of Standard Navy Colored Flares," Proceedings, Pyrotechnics and Explosives Applications Section, American Defense Preparedness Association, Fort Worth, Texas, September, 1983.

generally be counted upon to produce significant amounts of gas. Organic binders and sulfur should be avoided if a "gasless" composition is desired.

6. *Efficiency*: For a particular composition to be of practical interest, it must produce a significant amount of pyrotechnic effect *per gram* of mixture. Efficiency per unit *volume* is also an important consideration when available space is limited.
7. *Ignitibility*: A pyrotechnic composition must be capable of undergoing reliable ignition, and yet be stable in transportation and storage. The ignition behavior of every mixture must be studied, and the proper ignition system can then be specified for each. For easily-ignited materials, the "spit" from a burning black powder fuse is often sufficient. Another common igniter is a "squib" or electric match, consisting of a metal wire coated with a small dab of heat-sensitive composition. An electric current is passed through the wire, producing sufficient heat to ignite the squib. The burst of flame then ignites the main charge. For pyrotechnic mixtures with high ignition temperatures, a *primer* or *first fire* is often used. This is an easily-ignited composition that can be activated by a fuse or squib. The flame and hot residue produced are used to ignite the principal material. This topic will be treated in more detail in Chapter 5.

To produce the desired pyrotechnic effect from a given mixture, the chemist must be aware of the large number of variables that can affect performance. These factors must be held constant from batch to batch and day to day to achieve reproducible behavior. Substantial deviations can result from variations in any of the following [2]:

1. *Moisture*: The best rule is to avoid the use of water in processing pyrotechnic compositions, and to avoid the use of all hygroscopic (water-attracting) ingredients. If water *is* used to aid in binding and granulating, an efficient drying procedure must be included in the manufacturing process. The final product should be analyzed for moisture content, if reproducible burning behavior is critical.
2. *Particle size of ingredients*: Homogeneity, and pyrotechnic performance, will increase as the particle size of the various components is decreased. The finer the particle

TABLE 4.3 Effect of Particle Size on Performance of a Flare Composition[a]

Composition:	Component	% by weight	Average particle size
	Magnesium metal	48	see table below
	Sodium nitrate, $NaNO_3$	42	34 micrometers (10^{-6} meters)
	Laminac binder	8	-
	Polyvinyl chloride	2	27 micrometers

Magnesium average particle size, micrometers	Flare candlepower (1,000 candles)	Flare burning rate, inches/minute
437	130	2.62
322	154	3.01
168	293	5.66
110	285	5.84

[a]Reference 2.

size, the more reactive a particular composition should be, with all other factors held constant. Table 4.3 illustrates this principle for a sodium nitrate/magnesium flare composition. Note the similarity in performance for the two smallest particle sizes, suggesting that an upper performance limit may exist.

3. *Surface area of the reactants*: For a high-energy reaction to rapidly proceed, the oxidizer must be in intimate contact with the fuel. Decreasing particle size will increase this contact, as will increasing the available surface area of the particles. A smooth sphere will possess the *minimum* surface area for a given mass of material. An uneven, porous particle will exhibit much more free surface, and consequently will be a much more reactive material. Particle

TABLE 4.4 Effect of Particle Size on the Burning Rate of Tungsten Delay Mixtures[a]

	Mix A ("M 10")	Mix B ("ND 3499")
% Tungsten, W	40	38
% Barium chromate, $BaCrO_4$	51.8	52
% Potassium perchlorate, $KClO_4$	4.8	4.8
% Diatomaceous earth	3.4	5.2
Tungsten surface area, cm^2/gram	1377	709
Tungsten average diameter, 10^{-6} m	2.3	4.9
Burning rate of mixture, in/sec	0.24	0.046

[a]Reference 2.

size is important, but surface area can be even more critical in determining reactivity. Several examples of this phenomenon are presented in Tables 4.4 and 4.5.

4. *Conductivity*: For a column of pyrotechnic composition to burn smoothly, the reaction zone must readily travel down the length of the composition. Heat is transferred from layer to layer, raising the adjacent material to the ignition temperature of the particular composition. Good thermal conductivity can be essential for smooth propagation of burning, and this is an important role played by *metals* in many mixtures. Metals are the best thermal conductors, with organic compounds ranking among the worst. Table 2.10 lists the thermal conductivity values of some common materials.
5. *Outside container material*: Performance of a pyrotechnic mixture can be affected to a substantial extent by the type of material used to contain the mixed composition. If a good thermal conductor, such as a metal, is used, heat may be carried away from the composition through the wall of the container to the surroundings. The thickness

TABLE 4.5 Effect of Particle Size on Burning Rate[a]

Composition:		
	Titanium metal	48% by weight
	Strontium nitrate	45
	Linseed oil	4
	Chlorinated rubber	3

Titanium size range, micrometers	Relative burning rate
less than 6	1.00 (fastest)
6-10	0.68
10-14	0.63
14-18	0.50
greater than 18	0.37 (slowest)

Note: Curiously, the system showed the *opposite* effect for strontium nitrate. Decreasing the particle size of the oxidizer from 10.5 to 5.6 micrometers produced a 25% decrease in burning rate.
[a]Reference 5.

of such a metal wall will also be an important consideration. If sufficient heat does not pass down the length of the pyrotechnic mixture, burning may not propagate and the device will not burn completely. Organic materials, such as cardboard, are widely used to contain low-energy pyrotechnic compositions – such as highway fuses and fireworks – to minimize this problem (cardboard is a poor thermal conductor).

6. *Loading pressure*: There are two general rules to describe the effect of loading pressure on the burning behavior of a pyrotechnic composition. If the pyrotechnic reaction, in the post-ignition phase, is propagated via hot gases, then too high of a loading pressure will retard the passage of these hot gases down the column of composition. A lower rate, in units of grams of composition reacting per second, will be observed at high loading pressures. (*Note*: One must be cautious in interpreting burn rate data, because

TABLE 4.6 Effect of Loading Pressure on the Burning Rate of a Delay Mixture

Composition:	Barium chromate, $BaCrO_4$	90
	Boron	10

Loading pressure (1000 psi)	Burning rate (seconds/gram)[a]
36	.272 (fastest)
18	.276
9	.280
3.6	.287
1.3	.297
0.5	.309 (slowest)

Note: This is a "gasless" delay mixture - the burning rate *increases* as loading pressure increases. "Gassy" mixtures will show the opposite behavior.
[a]Reference 2.

an increase in loading pressure usually leads to an increase in the *density* of the composition. What may appear to be a *slower* rate, expressed in units of millimeters/second, may actually be a *faster* rate in terms of grams/second.)

If the propagation of the pyrotechnic reaction is a solid-solid or solid-liquid phenomenon, without the significant involvement of gas-phase components, then an increase in loading pressure should lead to an increase in burn rate (in grams per second). An example of this possibility is given in Table 4.6.

7. *Degree of confinement*: In Chapter 1, the variation in the burning behavior of black powder was discussed as a function of the degree of confinement. Increased confinement leads to accelerated burning. Shimizu reports a burning rate in air of .03-.05 meters/second for black powder paste impregnated in twine. The same material, enclosed in a

paper tube of one cm inside diameter, had a burning rate of 4.6-16.7 meters/second – over 100 times faster [4]! This behavior is typical of loose powders, and points out the potential danger of confining mixtures that burn quite sluggishly in the open air.

This effect is particularly important when consideration is given to the storage of pyrotechnic compositions. Containers and storage facilities should be designed to instantly *vent* in the case of pressure buildup. Such venting can quite effectively prevent many fires from progressing to explosions.

Two factors contribute to the effect of confinement on burning rate. First, as was discussed in Chapter 2, an increase in temperature produces an exponential increase in rate of a chemical reaction. In a confined high-energy system, the temperature of the reactants can rise dramatically upon ignition, as heat is not effectively lost to the surroundings. A sharp rise in reaction rate occurs, liberating more heat, raising the temperature further, accelerating the reaction until an explosion occurs or the reactants are consumed. The minimum quantity of material needed to produce an explosion, under a specified set of conditions, is referred to as the *critical mass*. Also, in a confined system, the hot gases that are produced can build up substantial pressure, driving the gases into the high-energy mixture and causing a rate acceleration.

Burning behavior can therefore be summarized in two words: *homogeneity* and *confinement*. An increase in either should lead to an increase in burning rate for most high-energy mixtures. Note, however, that "gasless" compositions do not show the dramatic confinement effects found for "gassy" compositions.

REQUIREMENTS FOR A GOOD HIGH-ENERGY MIXTURE

The requirements for a commercially-feasible high-energy mixture can be summarized as follows, keeping in mind the preceding discussions of materials and factors that affect performance:

1. The composition produces the desired effect and is efficient both in terms of effect/gram and effect/dollar.

2. The composition can easily and safely be manufactured, handled, transported, stored, and used, assuming normal treatment and the expected variations in temperature.
3. Storage lifetime is acceptable, even in humid conditions, and there is reasonably low toxicity associated with both the starting materials and reaction products.

These requirements seem rather simple, but they do restrict or eliminate a number of potential starting materials. These compounds must either be deleted from our "acceptable" list or special precautions must be taken in order to use them. Examples include:

Potassium dichromate ($K_2Cr_2O_7$): This is a strong oxidizer, but it only contains 16% oxygen by weight. It has a corrosive effect on the mucous membranes, and its toxicity and suspected carcinogenicity suggest the use of alternate oxidizers.

Ammonium perchlorate (NH_4ClO_4): This is a good oxidizer, and can be used to make excellent propellants and colored flames. However, it is a *self-contained* oxidizer-fuel system (much like ammonium nitrate). The mixing of NH_4^+ (fuel) and ClO_4^- (oxidizer) occurs at the ionic level. The potential for an explosion cannot be ignored. Conclusion: if this material is used, it must be treated with respect and minimum quantities of bulk powder should be prepared.

Magnesium metal (Mg): This is an excellent fuel and produces brilliant illuminating mixtures. The metal is water-reactive however, suggesting short shelf-life and possible spontaneous ignition if magnesium-containing mixtures become wet. Conclusion: replace magnesium with the more stable aluminum (or possibly titanium) metals. If magnesium gives the best effect, coat the metal with an organic, water-repelling material.

PREPARATION OF HIGH-ENERGY MIXTURES

The most hazardous operations in the high-energy chemistry field involve the mixing of oxidizer and fuel in large quantities, and the subsequent drying of the composition (if water or other liquid is used in the mixing and granulating processes). In these operations, large quantities of bulk powder are present in one location, and if accidental ignition should occur, there is a good chance that an explosive reaction rate may be reached.

For this reason, mixing and drying operations should be isolated from all other plant processes, and remote control equipment should be used wherever and whenever possible. All high-energy manufacturing facilities should be designed with the idea in mind that an accident *will* occur at some time during the life of the facility. The plant should be designed to minimize any damage to the facility, to the neighborhood, and most importantly, to the operating personnel.

The manufacturing operation can be divided into several stages:

1. *Preparation of the individual components*: Materials to be used in the manufacturing process may have to be dried, as well as ground or crushed to achieve the proper particle size, or screened to separate out large particles or foreign objects. Oxidizers should never be processed with the same equipment used for fuels, nor should oxidizers and fuels be stored in the same area prior to use. All materials *must* be clearly labeled at all times.

2. *Preparation of compositions*: This step is the key to proper performance. The more homogeneous a mixture is, the greater its reactivity will be. The high-energy chemist is always walking a narrow line in this area, however. By maximizing reactivity – with small particle sizes and intimate mixing – you are also increasing the chance of accidental ignition during manufacturing and storage. A compromise is usually reached, obtaining a material that performs satisfactorily but is reasonably safe to work with. This compromise is reached by careful specification of particle size, purity of starting materials, and safe operating procedures.

A variety of methods can be used for mixing. Materials can be blended through wire screens, using brushes. Hand-screening is still used in the fireworks industry, but should never be used with explosive or unstable mixtures. Brushes provide a safer method of screening the oxidizer and fuel together. Materials can also be tumbled together to achieve homogeneity, and this can (and should) be done remotely. Remote mixing is *strongly* recommended for sensitive explosive compositions such as the "flash and sound" powder used in firecrackers and salutes and the photoflash powders used by the military.

3. *Granulation*: Following mixing, the powders are often granulated, generally using a small percentage of binder to aid

in the process. The composition is treated with water or an organic liquid (such as alcohol), and then worked through a large-mesh screen. Grains of well-mixed composition are produced which will retain the homogeneity of the composition better than loose powder. Without the granulation step, light and dense materials might segregate during transportation and storage. The granulated material is dried in a remote, isolated area, and is then ready to be loaded into finished items. *Remember*: Sizable quantities of bulk powder are present at this stage, and the material must be protected from heat, friction, shock, and static spark.

4. *Loading*: An operator, working with the minimum quantity of bulk powder, loads the composition into tubes or other containers, or produces pellets for later use in finished items. The making of "stars" – small pieces of color-producing composition used in aerial fireworks – is an example of this pelleting operation.

5. *Testing*: An important final step in the manufacturing process is the continual testing of each lot of finished items to ensure proper performance. Significant differences in performance can be obtained by slight variation in the particle size or purity of any of the starting materials, and a *regular* testing program is the only way to be certain that proper performance is being achieved.

REFERENCES

1. T. L. Davis, *The Chemistry of Powder and Explosives*, John Wiley & Sons, Inc., New York, 1941.
2. U.S. Army Material Command, Engineering Design Handbook, Military Pyrotechnic Series, Part One, "Theory and Application," Washington, D.C., 1967 (AMC Pamphlet 706-185).
3. A. A. Shidlovskiy, *Principles of Pyrotechnics*, 3rd Ed., Moscow, 1964. (Translated by Foreign Technology Division, Wright-Patterson Air Force Base, Ohio, 1974.)
4. T. Shimizu, *Fireworks from a Physical Standpoint*, Part One, (trans. by A. Schuman), Pyrotechnica Publications, Austin, Texas, 1982.
5. B. J. Thomson and A. M. Wild, "Factors Affecting the Rate of Burning of a Titantium - Strontium Nitrate Based Composition," Proceeding of Pyrochem International 1975, Pyrotechnics Branch, Royal Armament Research and Development Establishment, United Kingdom, July, 1975.

A magnesium-containing flare burns with a brilliant white flame in the test tunnel of the Applied Sciences Department, Naval Weapons Support Center, Crane, Indiana. Special instrumentation can measure the intensity of the light output as a function of wavelength. "White light" compositions emit throughout the visible region of the electromagnetic spectrum (380-780 nanometers), with emission extending into the infrared and ultraviolet regions. Researchers at the Crane facility have performed extensive research on the theory and performance of illuminating flares, especially the sodium nitrate/magnesium system. (NWSC, Crane, Indiana)

5
IGNITION AND PROPAGATION

IGNITION PRINCIPLES

Successful performance of a high-energy mixture depends upon:

1. The ability to *ignite* the material using an external stimulus, as well as the *stability* of the composition in the absence of the stimulus.
2. The ability of the mixture, once ignited, to *sustain* burning through the remainder of the composition.

Therefore, a composition is required that *will* readily ignite and burn, producing the desired effect upon demand, while remaining quite stable during manufacture and storage. This is *not* an easy requirement to meet, and is one of the main reasons why a relatively small number of materials are used in high-energy mixtures.

For ignition to occur, a portion of the mixture must be heated to its *ignition temperature*, which is defined as the minimum temperature required for the initiation of a self-propagating reaction. Upon ignition, the reaction then proceeds on its own, in the absence of any additional energy input.

Application of the ignition stimulus (such as a spark or flame) initiates a complex sequence of events in the composition. The solid components may undergo crystalline phase transitions, melting, boiling, and decomposition. Liquid and vapor phases may be formed, and a chemical reaction will eventually occur at the surface where the energy input is applied, *if* the necessary *activation energy* has been provided.

The heat released by the occurrence of the high-energy reaction raises the temperature of the next layer of composition. If the heat evolution and thermal conductivity are sufficient to supply the required activation energy to this next layer, further reaction will occur, liberating additional heat and propagation of the reaction down the length of the column of mixture takes place. The rates, and quantity, of heat transfer *to*, heat production *in*, and heat loss *from* the high-energy composition are all critical factors in achieving propagation of burning and a self-sustaining chemical reaction.

The combustion process itself is quite complex, involving high temperatures and a variety of short-lived, high-energy chemical species. The solid, liquid, and vapor states may all be present in the actual flame, as well as in the region immediately adjacent to it. *Products* will be formed as the reaction proceeds, and they will either escape as gaseous species or accumulate as solids in the reaction zone (Figure 5.1).

A *moving*, high-temperature reaction zone, progressing through the composition, is characteristic of a combustion (or "burning") reaction. This zone separates unreacted starting material from the reaction products. In "normal" chemical reactions, such as those carried out in a flask or beaker, the entire system is at the same temperature and molecules react randomly throughout the container. Combustion is distinguished from detonation by the absence of a pressure differential between the region undergoing reaction and the remainder of the unreacted composition [1].

A variety of factors affect the ignition temperature and the burning rate of a high-energy mixture, and the chemist has the ability to alter most of these factors to achieve a desired change in performance.

One requirement for ignition appears to be the need for either the oxidizer or fuel to be in the liquid (or vapor) state, and reactivity becomes even more certain when *both* are liquids. The presence of a low-melting fuel can substantially lower the ignition temperature of many compositions [2]. Sulfur and organic compounds have been employed as "tinders" in high-energy mixtures to facilitate ignition. Sulfur melts at 119°C, while most sugars, gums, starches, and other organic polymers have melting points or decomposition temperatures of 300°C or less (Table 5.1).

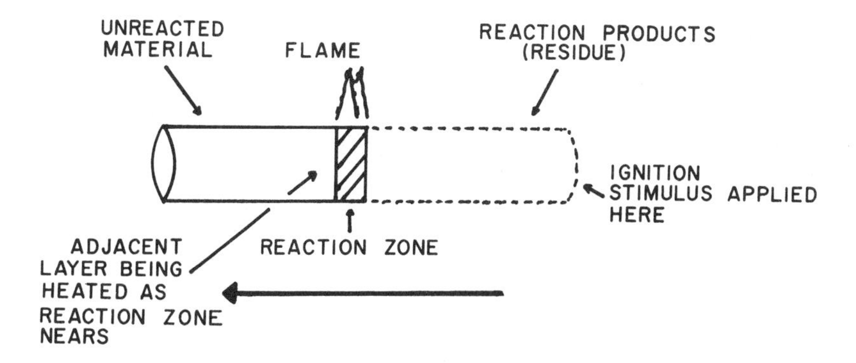

FIG. 5.1 Burning pyrotechnic composition. Several major regions are present in a reacting pyrotechnic composition. The actual self-propagating exothermic process is occurring in the reaction zone. High temperature, flame and smoke production, and the likely presence of gaseous and liquid materials characterize this region. Behind the advancing reaction zone are solid products formed during the reaction (unless all products were gaseous). Immediately ahead of the reaction zone is the next layer of composition that will undergo reaction. This layer is being heated by the approaching reaction, and melting, solid-solid phase transitions, and low-velocity pre-ignition reactions may be occurring. The thermal conductivity of the composition is quite important in transferring heat from the reaction zone to the adjacent, unreacted material. Hot gases as well as hot solid and liquid particles aid in the propagation of burning.

The oxidizers used in high-energy mixtures are generally ionic solids, and the "looseness" of the ionic lattice is quite important in determining their reactivity [3]. A crystalline lattice has some vibrational motion at normal room temperature, and the amplitude of this vibration increases as the temperature of the solid is raised. At the melting point, the forces holding the crystalline solid together collapse, producing the randomly-oriented liquid state. For reaction to occur in a high-energy system, the fuel and oxygen-rich oxidizer anion must become intimately mixed, on the ionic or molecular level. Liquid fuel can diffuse into the solid oxidizer lattice *if* the vibrational amplitude in the crystal is sufficient.

TABLE 5.1 Effect of Sulfur and Organic Fuels on Ignition Temperature

Composition		(% by weight)	Ignition temperature, °C
IA.	$KClO_4$	66.7	446[a]
	Al	33.3	
IB.	$KClO_4$	64	360
	Al	22.5	
	S	10	
	Sb_2S_3	3.5	
IIA.	$BaCrO_4$	90	615[a] (3.1 ml per gram of evolved gas)
	B	10	
IIB.	$BaCrO_4$	90	560 (29.5 ml per gram of evolved gas)
	B	10	
	Vinyl alcohol/ acetate resin	1 (additional %)	
IIIA.	$NaNO_3$	50	772[b] (50 mg sample, heated at 50°C/min.)
	Ti	50	
IIIB.	$NaNO_3$	50	357
	Ti	50	
	Boiled linseed oil	6 (additional %)	

[a]Reference 10.
[b]Reference 2.

Once sufficient heat is generated to begin decomposing the oxidizer, the higher-temperature combustion reaction begins, involving free oxygen gas and very rapid rates. We are concerned here with the processes that *initiate* the ignition process.

Professor G. Tammann, one of the pioneers of solid-state chemistry, considered the importance of lattice motion to reactivity, and used the ratio of the actual temperature of a solid divided by the melting point of the solid (on the Kelvin or "absolute" scale) to quantify this concept.

$$\alpha = T(\text{solid})/T(\text{melting point}) \text{ (in K)} \quad (5.1)$$

Tammann proposed that diffusion of a mobile species into a crystalline lattice should be "significant" at an α-value of 0.5 (or halfway to the melting point, on the Kelvin scale). At this temperature, later termed the *Tammann temperature*, a solid has approximately 70% of the vibrational freedom present at the melting point, and diffusion into the lattice becomes probable [3]. If this is the approximate temperature where diffusion becomes probable, it is therefore also the temperature where a chemical reaction between a good oxidizer and a mobile, reactive fuel becomes possible. This is a very important point from a safety standpoint – the potential for a reaction may exist at surprisingly low temperatures, especially with sulfur or organic fuels present. Table 5.2 lists the Tammann temperatures of some of the common oxidizers. The *low* temperatures shown for potassium chlorate and potassium nitrate may well account for the large number of mysterious, accidental ignitions that have occurred with compositions containing these materials.

Ease of initiation also depends upon the particle size and surface area of the ingredients. This factor is especially important for the metallic fuels with melting point higher than or comparable to that of the oxidizer. Some metals -- including aluminum, magnesium, titanium, and zirconium -- can be quite hazardous when present in fine particle size (in the 1-5 micrometer range). Particles this fine may spontaneously ignite in air, and are quite sensitive to static discharge [4]. For safety reasons, reactivity is sacrificed to some extent when metal powders are part of a mixture, and larger particle sizes are used to minimize accidental ignition.

Several examples will be given to illustrate these principles. In the potassium nitrate/sulfur system, the liquid state initially appears during heating with the melting of sulfur at 119°C. Sulfur occurs in nature as an 8-member ring – the S_8 molecule. This ring begins to fragment into species such as S_3 at temperatures above 140°C. However, even with these fragments present, reaction between sulfur and the solid KNO_3 does not occur at a rate sufficient to produce ignition until the KNO_3 melts at 334°C. Intimate mixing can occur when both species are in the liquid state, and ignition is observed just above the KNO_3 melting point. Although some reaction presumably occurs between sulfur and

TABLE 5.2 Tammann Temperatures of the Common Oxidizers

Oxidizer	Formula	Melting point, °C	Melting point, °K	Tammann temperature, °C
Sodium nitrate	$NaNO_3$	307	580	17
Potassium nitrate	KNO_3	334	607	31
Potassium chlorate	$KClO_3$	356	629	42
Strontium nitrate	$Sr(NO_3)_2$	570	843	149
Barium nitrate	$Ba(NO_3)_2$	592	865	160
Potassium perchlorate	$KClO_4$	610	883	168
Lead chromate	$PbCrO_4$	844	1117	286
Iron oxide	Fe_2O_3	1565	1838	646

solid KNO_3 below the melting point, the low heat output obtained from the oxidation of sulfur combined with the *endothermic* decomposition of KNO_3 prevent ignition from taking place until the entire system is liquid. Only then is the reaction rate great enough to produce a self-propagating reaction. Figures 5.2-5.4 show the thermograms of the components and the mixture. Note the strong exotherm corresponding to ignition for the KNO_3/S mixture.

In the potassium chlorate/sulfur system, a different result is observed. Sulfur again melts at 119°C and begins to fragment above 140°C, but a strong exotherm corresponding to ignition of the composition is found well below 200°C! Potassium chlorate has a melting point of 356°C, so ignition is taking place well below the melting point of the oxidizer. We recall, though, that $KClO_3$ has a Tammann temperature of 42°C. A mobile species – such as liquid, fragmented sulfur – can penetrate the lattice well below the melting point and be in position to react. We also recall that the thermal decomposition of $KClO_3$ is *exothermic* (10.6 kcal of heat is evolved per mole of oxidizer that decomposes). A compounding of heat evolution is obtained – heat is released by the $KClO_3/S$ reaction *and* by the decomposition of additional $KClO_3$,

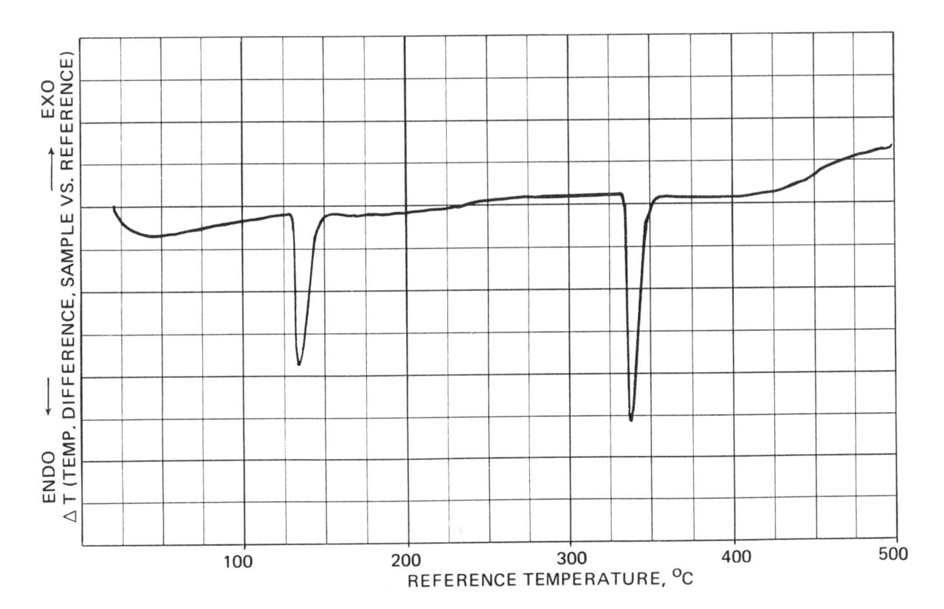

FIG. 5.2 Thermogram of pure potassium nitrate. Endotherms are observed near 130° and 334°C. These peaks correspond to a rhombic-to-trigonal crystalline phase transition and melting. Note the sharpness of the melting point endotherm near 334°C. Pure compounds will normally melt over a very narrow range. Impure compounds will have a broad melting point endotherm.

generating oxygen to react with additional sulfur. More heat is generated and an Arrhenius-type rate acceleration occurs, leading to ignition well below the melting point of the oxidizer. This combination of low Tammann temperature and exothermic decomposition helps account for the dangerous and unpredictable nature of potassium chlorate. Figures 5.5-5.6 show the thermal behavior of the $KClO_3$/S system.

As we proceed to higher-melting fuels and oxidizers, we see a corresponding increase in the ignition temperatures of two-component mixtures containing these materials. The lowest ignition temperatures are associated with combinations of low-melting fuels and low-melting oxidizers, while high-melting combinations generally display higher ignition points. Table 5.3 gives some examples of this principle.

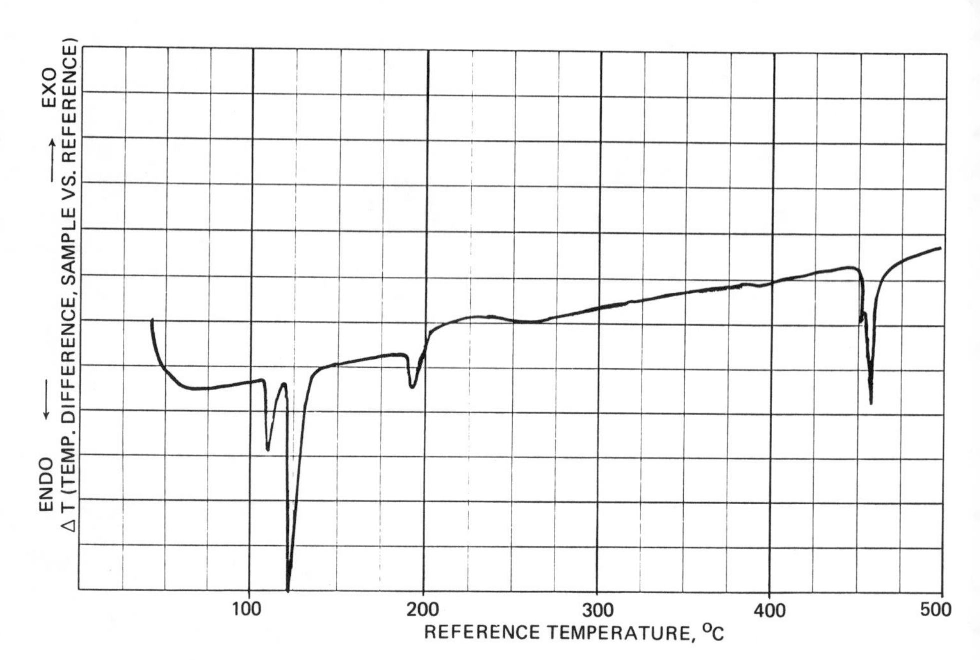

FIG. 5.3 A sulfur thermogram. Endotherms for a rhombic-to-monoclinic crystalline phase transition and melting are seen at 105° and 119°C, respectively. An additional endotherm is observed near 180°. This peak corresponds to the fragmentation of liquid S_8 molecules into smaller units. Finally, vaporization is observed near 450°C.

Table 5.3 shows that several potassium nitrate mixtures with low-melting fuels have ignition temperatures near the 334°C melting point of the oxidizer. Mixtures of KNO_3 with higher-melting metal fuels show substantially higher ignition temperatures. Table 5.4 shows that a variety of magnesium-containing compositions have ignition temperatures close to the 649°C melting point of the metal.

A problem with trying to develop logical theory using literature values of ignition temperatures is the substantial variation in observed values that can occur depending upon the experimental conditions employed to measure the ignition points. Ratio

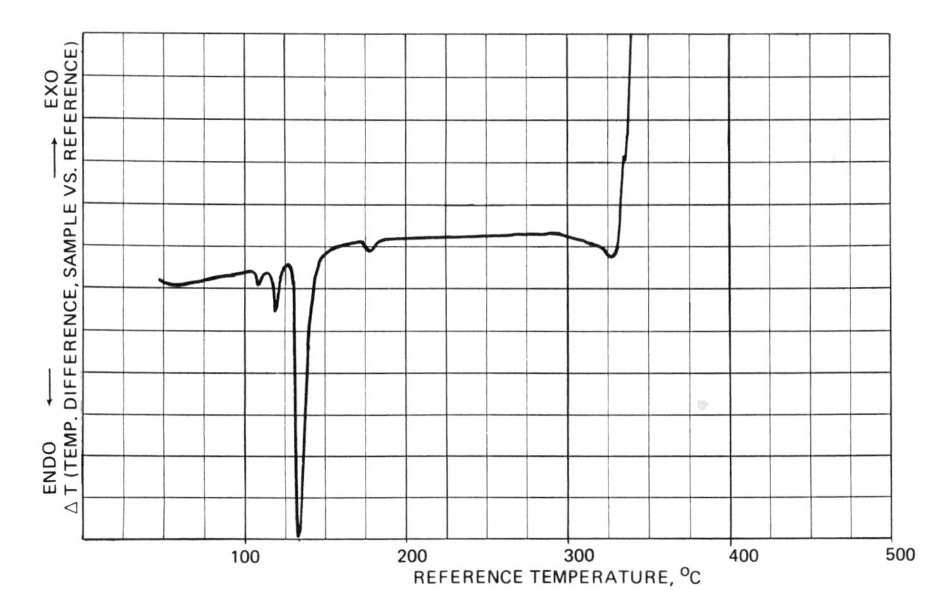

FIG. 5.4 The potassium nitrate/sulfur/aluminum system. Endotherms for sulfur can be seen near 105° and 119°C, followed by the potassium nitrate phase transition near 130°C. As the melting point of potassium nitrate is approached (334°C), an exotherm is observed. A reaction has occurred between the oxidizer and fuel, and ignition of the mixture evolves a substantial amount of heat.

of components, degree of mixing, loading pressure (if any), heating rate, and quantity of sample can all influence the observed ignition temperature.

The traditional method for measuring ignition temperatures, used extensively by Henkin and McGill in their classic studies of the ignition of explosives [6], consists of placing small quantities (3 or 25 milligrams, depending on whether the material detonates or deflagrates) of composition in a constant-temperature bath and measuring the time required for ignition to occur. *Ignition temperature* is defined, using this technique, as the temperature at which ignition will occur within five seconds. Data obtained in this type of study can be plotted to yield interesting information, as shown in Figure 5.7.

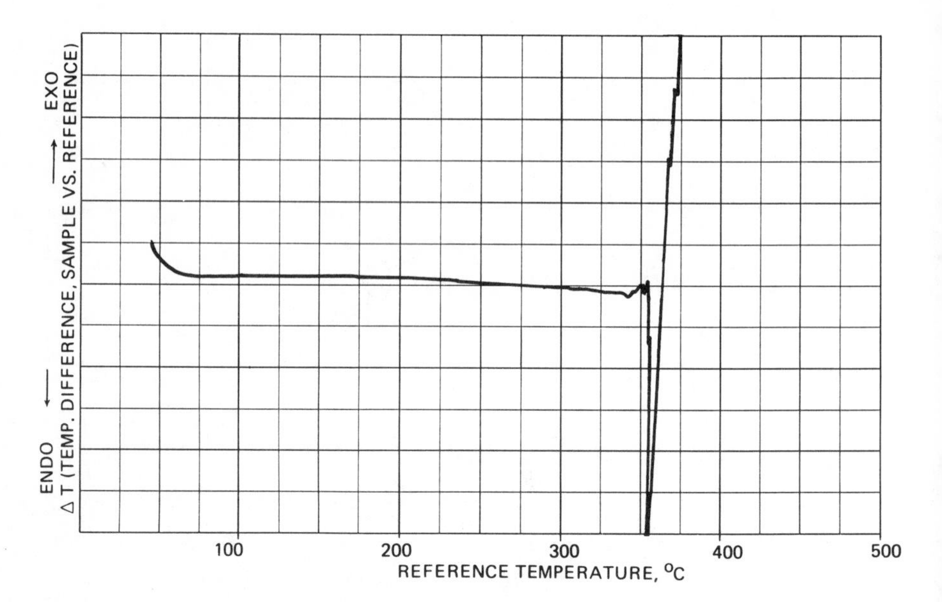

FIG. 5.5 Thermogram of pure potassium chlorate, $KClO_3$. No thermal events are observed prior to the melting point (356°C). Exothermic decomposition occurs above the melting point as oxygen gas is liberated.

Data from time versus temperature studies can also be plotted as log time vs. 1/T, yielding straight lines as predicted by the Arrhenius Equation (eq. 2.4). Figure 5.8 illustrates this concept, using the same data plotted in Figure 5.7. Activation energies can be obtained from such plots. Deviations from linear behavior and abrupt changes in slope are sometimes observed in Arrhenius plots due to changes in reaction mechanism or other complex factors.

"Henkin-McGill" plots can be quite useful in the study of ignition, providing us with important data on temperatures at which spontaneous ignition will occur. These data can be especially useful in estimating maximum storage temperatures for high-energy compositions – the temperature should be one corresponding to *infinite* time to ignition (below the "spontaneous ignition temperature," minimum – S.I.T (min) – shown in Figure 5.7). At any temperature above this point, ignition during storage is possible.

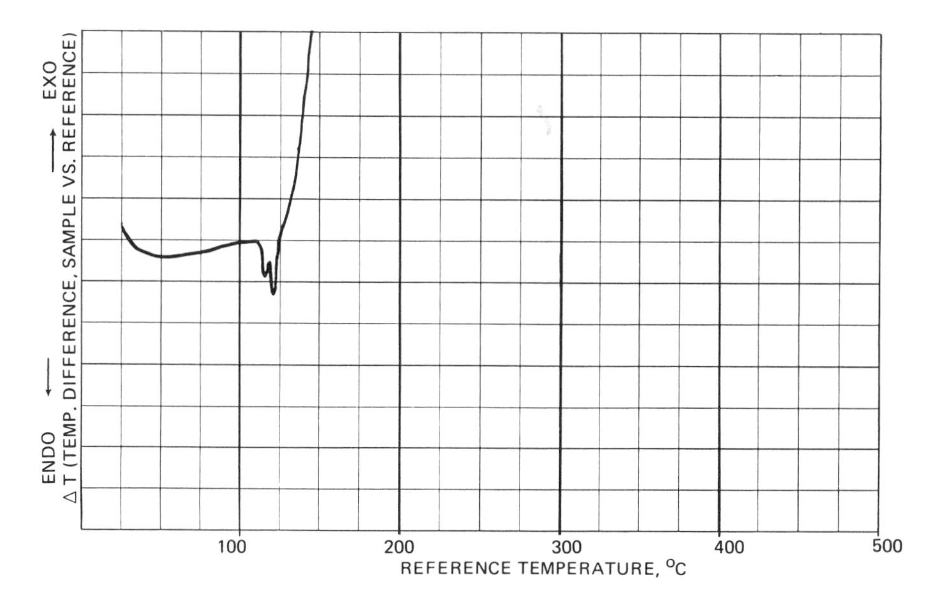

FIG. 5.6 The potassium chlorate/sulfur system. Sulfur endotherms are seen near 105° and 119°C, as expected. A violent exothermic reaction is observed below 150°C. The ignition temperature is approximately 200 degrees below the melting point of the oxidizer ($KClO_3$ m.p. = 356°C). Ignition occurs near the temperature at which S_8 molecules fragment into smaller units.

Ignition temperatures can also be determined by differential thermal analysis (DTA), and these values usually correspond well to those obtained by a Henkin-McGill study. Differences in heating rate can cause some variation in values obtained with this technique. For *any* direct comparison of ignition temperatures, it is best to run all of the mixtures of interest under identical experimental conditions, thereby minimizing the number of variables.

One must also keep in mind that these experiments are measuring the *temperature sensitivity* of a particular composition, in which the entire sample is heated to the experimental temperature. Ignition sensitivity can also be discussed in terms of the relative ease of ignition due to other types of potential stimuli, including static spark, impact, friction, and flame.

TABLE 5.3 Ignition Temperatures of Pyrotechnic Mixtures

Component[a]		Melting point, °C	Ignition temperature, °C
I.	$KClO_3$	356	150
	S	119	
II.	$KClO_3$	356	195[b]
	Lactose	202	
III.	$KClO_3$	356	540[b]
	Mg	649	
IV.	KNO_3	334	390[b]
	Lactose	202	
V.	KNO_3	334	340
	S	119	
VI.	KNO_3	334	565[b]
	Mg	649	
VII.	$BaCrO_4$ (90)	Decomposition at high temperatures	685[c]
	B (10)	2300	

[a]Mixtures were in stoichiometric proportions unless otherwise indicated.
[b]Reference 1.
[c]Reference 4.

SENSITIVITY

Sensitivity of a high-energy mixture to an ignition stimulus is influenced by a number of factors. The *heat output* of the fuel is quite important, with sensitivity generally increasing as the fuel's heat of combustion increases. Mixes containing magnesium or aluminum metal, or charcoal, can be quite sensitive to static spark or a fire flash, while mixes containing sulfur as the lone fuel are usually less sensitive, due to the low heat output of sulfur. Ignition of a small quantity of material by static energy does not

TABLE 5.4 Ignition Temperatures of Magnesium-Containing Mixtures[a]

Oxidizer	Ignition temperature, °C[b]
$NaNO_3$	635
$Ba(NO_3)_2$	615
$Sr(NO_3)_2$	610
KNO_3	650
$KClO_4$	715

Note: All mixtures contain 50% magnesium by weight.
[a]Reference 5.
[b]Loading pressure was 10,000 psi.

liberate sufficient energy, in a sulfur mix, to generate a self-propagating process. A greater quantity of material must react at once to produce ignition.

Another important factor is the thermal stability and heat of decomposition of the oxidizer. Potassium chlorate mixtures tend to be much more sensitive to ignition than potassium nitrate compositions, due to the exothermic nature of the decomposition of $KClO_3$. Mixtures containing very stable oxidizers – such as ferric oxide (Fe_2O_3) and lead chromate ($PbCrO_4$) – can be quite difficult to ignite, and a more-sensitive composition frequently has to be used in conjunction with these materials to effect ignition.

A mixture of a good fuel (e.g., Mg) with an easily-decomposed oxidizer (e.g., $KClO_3$) should be quite sensitive to a variety of ignition stimuli. A composition with a poor fuel and a stable oxidizer should be much less sensitive, if it can be ignited at all! Ignition temperature, as determined by DTA or a Henkin-McGill study, is but one measure of sensitivity, and there is not any simple correlation between ignition temperature and static spark or friction sensitivity. Some mixtures with reasonably high ignition temperatures ($KClO_4$ and Al is a good example) can be

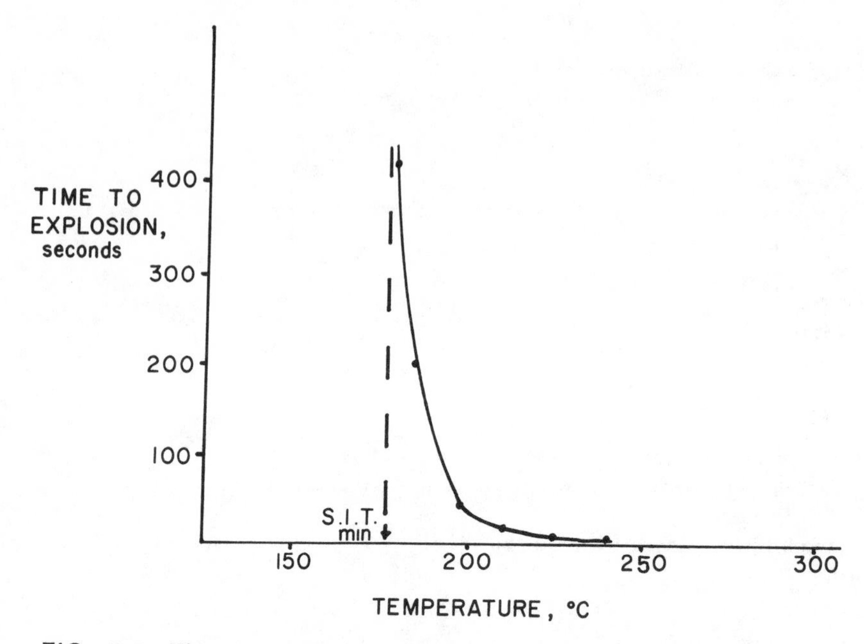

FIG. 5.7 Time to explosion versus temperature for nitrocellulose. As the temperature of the heating bath is raised, the time to explosion decreases exponentially, approaching an instantaneous value. The extrapolated temperature value corresponding to infinite time to explosion is called the spontaneous ignition temperature, minimum (S.I.T. min). Source of the data: reference 6.

quite spark sensitive, because the reaction is highly exothermic and becomes self-propagating once a small portion is ignited. *Sensitivity* and *output* are not necessarily related and are determined by different sets of factors. A given mixture can have high sensitivity and low output, low sensitivity and high output, etc. Those mixtures that have *both* high sensitivity and substantial output are the ones that must be treated with the greatest care. Potassium chlorate/sulfur/aluminum "flash and sound" mixture is an example of this type of dangerous composition.

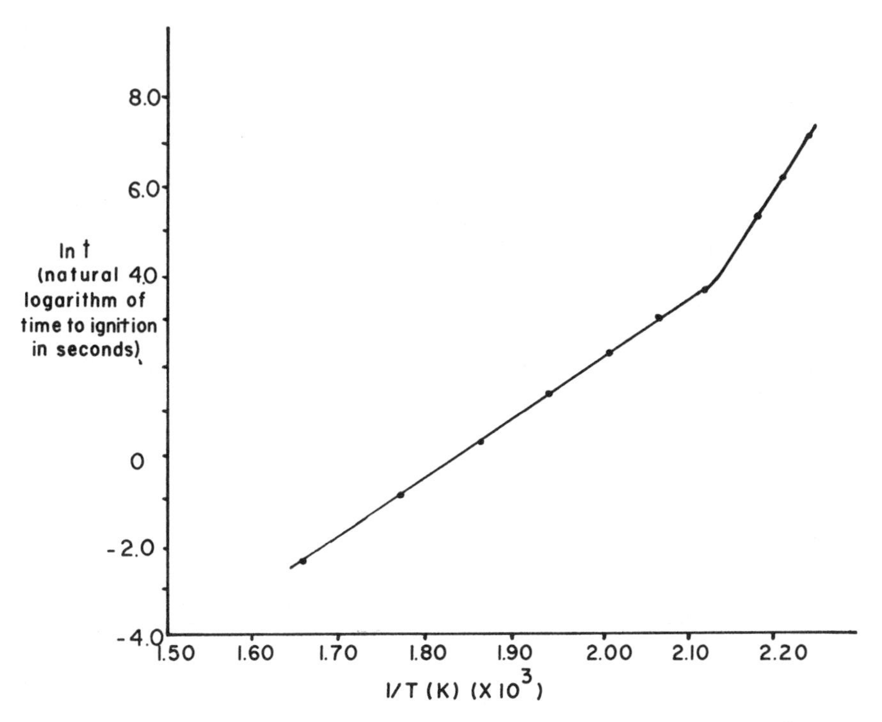

FIG. 5.8 "Henkin-McGill Plot" for nitrocellulose. The natural logarithm of the time to ignition is plotted versus the reciprocal of the absolute temperature (°K). A straight line is produced, and activation energies can be calculated from the slope of the line. The break in the plot near 2.1 may result from a change in the reaction mechanism at that temperature. Source: reference 6.

PROPAGATION OF BURNING

Factors

The ignition process initiates a self-propagating, high-temperature chemical reaction at the surface of the mixture. The rate at which the reaction then proceeds through the remainder of the composition will depend on the nature of the oxidizer and and fuel, as well as on a variety of other factors. "Rate" can be expressed in two ways – mass reacting per unit time or length burned per unit time. The loading pressure used, and the resulting density of the composition, will determine the relationship between these two rate expressions.

Reaction velocity is primarily determined by the selection of the oxidizer and fuel. The rate-determining step in many high-energy reactions appears to be an *endothermic* process, with decomposition of the oxidizer frequently the key step. The higher the decomposition temperature of the oxidizer, and the more endothermic the decomposition, the slower the burning rate will be (with all other factors held constant).

Shimizu reports the following reactivity sequence for the most-common of the fireworks oxidizers [8]:

$$KClO_3 > NH_4ClO_4 > KClO_4 > KNO_3$$

Shimizu notes that potassium nitrate is *not* slow when used in black powder and metal-containing compositions in which a "hot" fuel is present. Sodium nitrate is quite similar to potassium nitrate in reactivity.

Shidlovskiy has gathered data on burning rates for some of the common oxidizers [1]. Table 5.5 contains data for oxidizers with a variety of fuels. Again, note the high reactivity of potassium chlorate.

TABLE 5.5 Burning Rates of Stoichiometric Binary Mixtures[a]

	Linear burning rate, mm/sec[b]			
	Oxidizer			
Fuel	$KClO_3$	KNO_3	$NaNO_3$	$Ba(NO_3)_2$
Sulfur	2	X[c]	X	-
Charcoal	6	2	1	0.3
Sugar	2.5	1	0.5	0.1
Shellac	1	1	1	0.8

[a]Reference 1.
[b]Compositions were pressed in cardboard tubes of 16 mm diameter.
[c]X indicates that the mixture did not burn.

The fuel also plays an important role in determining the rate of combustion. Metal fuels, with their highly exothermic heats of combustion, tend to increase the rate of burning. The presence of low-melting, volatile fuels (sulfur, for example) tends to retard the burning rate. Heat is used up in melting and vaporizing these materials rather than going into raising the temperature of the adjacent layers of unreacted mixture and thereby accelerating the reaction rate. The presence of moisture can greatly retard the burning rate by absorbing substantial quantities of heat through vaporization. The heat of vaporization of water – 540 calories/gram at 100°C – is one of the largest values found for liquids. Benzene, C_6H_6, as an example, has a heat of vaporization of only 94 calories/gram at its boiling point, 80°C.

The higher the ignition temperature of a fuel, the slower is the burning rate of compositions containing the material, again with all other factors equal. Shidlovskiy notes that aluminum compositions are slower burning than corresponding magnesium mixtures due to this phenomenon [1].

The transfer of heat from the burning zone to the adjacent layers of unreacted composition is also critical to the combustion process. Metal fuels aid greatly here, due to their high thermal conductivity. For binary mixtures of oxidizer and fuel, combustion rate increases as the metal percentage increases, well past the stoichiometric point. For magnesium mixtures, this effect is observed up to 60-70% magnesium by weight. This behavior results from the increasing thermal conductivity of the composition with increasing metal percentage, and from the reaction of excess magnesium, vaporized by the heat evolved from the pyrotechnic process, with oxygen from the atmosphere [1].

Stoichiometric mixtures or those with an excess of a metallic fuel are typically the fastest burners. Sometimes it is difficult to predict exactly what the stoichiometric reaction(s) will be at the high reaction temperatures encountered with these systems, so a trial-and-error approach is often advisable. A series of mixtures should be prepared – varying the fuel percentage while keeping everything else constant. The percentage yielding the maximum burning rate is then experimentally determined.

Variation in loading density, achieved by varying the pressure used to consolidate the composition in a tube, can also affect the burning rate. A "typical" high-energy reaction evolves a substantial quantity of gaseous products and a significant portion of the actual combustion reaction occurs in the vapor phase. For these reactions, the combustion rate (measured in grams consumed/second) will increase as the loading density *decreases*. A loose powder should burn the fastest, perhaps reaching an explosive velocity, while a highly-consolidated mixture, loaded under considerable pressure, will burn much more slowly. The combustion front in such mixtures is carried along by hot gaseous products. The more *porous* the composition is, the faster the reaction should be. The "ideal" fast composition is one that has been granulated to achieve a high degree of homogeneity *within* each particle but yet consists of small grains of powder with high surface area. Burning will accelerate rapidly through a loose collection of such particles.

The exception to this "loading pressure rule" is the category of "gasless" compositions. Here, burning is believed to propagate through the mixture *without* the involvement of the vapor phase, and an increase in loading pressure should lead to an *increase* in burning rate, due to more efficient heat transfer via tightly compacted solid and liquid particles. Thermal conductivity is quite important in the burning rate of these compositions. Table 4.6 illustrates the effect of loading pressure for the "gasless" barium chromate/boron system.

Effect of External Pressure

The gas pressure (if any) generated by the combustion products, combined with the prevailing atmospheric pressure, will also affect the burning rate. The general rule here predicts that an increase in burning rate will occur as the external pressure increases. This factor can be especially important when oxygen is a significant component of the gaseous phase. The magnitude of the external pressure effect indicates the extent to which the vapor phase is involved in the combustion reaction.

The effect of external pressure on the burning rate of black powder has been quantitatively studied. Shidlovskiy reports the experimental empirical equation for the combustion of black powder to be

$$\underset{\substack{\uparrow \\ \text{(cm/sec)}}}{\text{burning rate}} = 1.21\ P\ (0.24) \qquad (5.2)$$

where P = pressure, in atmospheres. Predicted burning rates for black powder, calculated using this equation, are given in Table 5.6.

For "gasless" heat and delay compositions, little external pressure effect is expected. This result, plus the increase in burning rate observed with an increase in loading pressure, can be considered good evidence for the *absence* of any significant gas-phase involvement in a particular combustion mechanism.

TABLE 5.6 Predicted Burning Rates for Black Powder at Various External Pressures

External pressure, atm	External pressure, p.s.i.	Linear burning rate, cm/sec
1	14.7	1.21
2	29.4	1.43
5	73.5	1.78
10	147	2.10
15	221	2.32
20	294	2.48
30	441	2.71

Note: The Shidlovskiy equation is valid for the pressure range 2-30 atmospheres.

For the ferric oxide/aluminum (Fe_2O_3/Al), manganese dioxide/aluminum (MnO_2/Al), and chromic oxide/magnesium (Cr_2O_3/Mg) systems, slight gas phase involvement is indicated by the 3-4 fold rate increase observed as the external pressure is raised from 1 to 150 atm. The chromic oxide/aluminum system, however, reportedly burns at exactly the same rate — 2.4 millimeters/sec — at 1 and 100 atm; suggesting that it is a true "gasless" system [1].

Data for the burning rate of a delay system as a function of external pressure (a nitrogen atmosphere was used) are given in Table 5.7.

Another matter to consider is whether or not pyrotechnic compositions will burn, and at what rate, at very *low* pressures. For reactions that use oxygen from the air as an important part of their functioning, a substantial drop in performance is expected at low pressure. Mixtures high in fuel (such as the magnesium-rich illuminating compositions) will not burn well at low pressures. Stoichiometric mixtures — in which all the oxygen needed to burn the fuel is provided by the oxidizer — should be the least affected by pressure variations.

TABLE 5.7 Burning Rate of a Delay Mixture as a Function of External Pressure[a]

Composition: Potassium permanganate, $KMnO_4$ 64%
Antimony, Sb 36%

External pressure, p.s.i.	Burning rate[b], cm/sec
14.7	.202
30	.242
50	.267
80	.296
100	.310
150	.343
200	.372
300	.430
500	.501
800	.529
1100	.537
1400	.543

[a]Source: Glasby, J.S., "The Effect of Ambient Pressure on the Velocity of Propagation of Half-Second and Short Delay Compositions," Report No. D.4152, Imperial Chemical Industries, Nobel Division, Ardeer, Scotland.
[b]Compositions were loaded into 10.5 mm brass tubes at a loading pressure of 20,000 p.s.i.

Burning Surface Area

The burning rate – expressed either in grams/second or millimeters/second – will increase as the burning surface area increases. Small grains will burn faster than large grains due to their greater surface area per gram. Compositions loaded into a narrow tube should burn more slowly than the same material in a wide tube. The heat loss to the walls of the container is less significant for a wide-bore tube, relative to the heat retained by the composition. For each composition, and each loading pressure, there will be a minimum diameter capable of producing stable burning. This minimum diameter will decrease as the exothermicity of the composition increases.

A metal tube is particularly effective at *removing* heat from a burning composition, and propagation of burning down a narrow column can be difficult for all but the hottest of mixtures if metal is used for the container material. On the other hand, the use of a metal wire for the *center* of the popular wire "sparkler" retains the heat evolved by the barium nitrate/aluminum reaction and *aids* in propagating the burning down the length of thin pyrotechnic coating.

A mixture that burns well in a narrow tube may possibly reach an explosive velocity in a thicker column, so careful experiments should be done any time a diameter change is made. For narrow tubes, one must watch out for possible restriction of the tube by solid reaction products, thereby preventing the escape of gaseous products. An explosion may result if this occurs, especially for fast compositions.

External Temperature

Finally, with a knowledge of the Arrhenius rate-temperature relationship, it can be anticipated that burning rate will also depend on the initial temperature of the composition. Considerably more heat input is needed to provide the necessary activation energy at -30°C than is needed when the mixture is initially at +40°C (or higher). Hence, both ignition and burning rate will be affected by variations in external temperature; the effect should be most pronounced for compositions of low exothermicity and low flame temperature. For black powder, a 15% slower rate is reported at 0°C versus 100°C, at external pressure of one atm [1]. Some high explosives show an even greater temperature sensitivity. Nitroglycerine, for example, is 2.9 times faster at 100°C than it is at 0°C [1].

Combustion Temperature

A pyrotechnic reaction generates a substantial quantity of heat, and the actual flame temperature reached by these mixtures is an area of study that has been attacked from both the experimental and theoretical directions.

Flame temperatures can be measured directly, using special high-temperature optical methods. They can also be calculated (estimated) using heat of reaction data and thermochemical values for heat of fusion and vaporization, heat capacity, and transition temperatures. Calculated values tend to be higher than the actual experimental results, due to heat loss to the surroundings as well as the endothermic decomposition of some of the reaction products. Details regarding these calculations, with several examples, have been published [5].

Considerable heat will be used to melt and to vaporize the reaction products. Vaporization of a reaction product is commonly the limiting factor in determining the maximum flame temperature. For example, consider a beaker of water at 25°C. As the water is heated, at one atmosphere pressure, the temperature of the liquid rises rather quickly to a value of 100°C. To heat the water over this temperature range, a heat input of approximately 1 calorie per gram per degree rise in temperature is required. To raise 500 grams of water from 25° to 100°C will require

Heat required = (grams of water)(heat capacity)(°T change)

= (500 grams)(1 cal/deg-gram)(75 deg)

= 37,500 calories

Once the water reaches 100°C, however, the temperature increase stops. The water boils, as liquid is converted to the vapor state, and 540 calories of heat is needed to convert 1 gram of water from liquid to vapor. To vaporize 500 grams of water, at 100°C,

(500 grams)(540 cal/gram) = 270,000 calories

of heat is required. Until this quantity of heat is put into the system, and all of the water is vaporized, no further temperature increase will occur. Similar phenomena involving the vaporization of reaction products such as magnesium oxide (MgO) and aluminum oxide (Al_2O_3) tend to limit the temperature attained in pyrotechnic flames. The boiling points of some common combustion products are given in Table 5.8.

Mixtures using organic (carbon-containing) fuels usually attain lower flame temperatures than those compositions consisting of an oxidizer and a metallic fuel. This reduction in flame temperature can be attributed to the lower heat output of the organic fuels versus metals, as well as to some heat consumption going towards the decomposition and vaporization of the organic fuel and its by-products. The presence of even small quantities of organic components can markedly lower the flame temperature, as the available oxygen is consumed by the carbonaceous material

TABLE 5.8 Melting and Boiling Points of Common Non-Gaseous Pyrotechnic Products[a]

Compound	Formula	Melting point, °C	Boiling point, °C
Aluminum oxide	Al_2O_3	2072	2980
Barium oxide	BaO	1918	ca. 2000
Boron oxide	B_2O_3	450	ca. 1860
Magnesium oxide	MgO	2852	3600
Potassium chloride	KCl	770	1500 (sublimes)
Potassium oxide	K_2O	350 (decomposes)	-
Silicon dioxide	SiO_2	1610 (quartz)	2230
Sodium chloride	$NaCl$	801	1413
Sodium oxide	Na_2O	1275 (sublimes)	-
Strontium oxide	SrO	2430	ca. 3000
Titanium dioxide	TiO_2	1830-1850 (rutile)	2500-3000
Zirconium dioxide	ZrO_2	ca. 2700	ca. 5000

[a]Source: R. C. Weast (ed.), CRC Handbook of Chemistry and Physics, 63rd ed., CRC Press, Inc., Boca Raton, Florida, 1982.

rather than metallic fuel [7]. Table 5.9 illustrates this behavior, with data reported by Shimizu [8].

This reduction of flame temperature can be minimized somewhat by using binders with as high an oxygen content as possible. In such binders, the carbon atoms are already partially oxidized, and they will therefore consume less oxygen in going to carbon dioxide during the combustion process. The balanced chemical equations for the combustion of hexane (C_6H_{14}) and glucose ($C_6H_{12}O_6$) illustrate this (both are six-carbon molecules):

$$C_6H_{14} + 9.5\ O_2 \rightarrow 6\ CO_2 + 7\ H_2O$$

$$C_6H_{12}O_6 + 6\ O_2 \rightarrow 6\ CO_2 + 6\ H_2O$$

TABLE 5.9 Effect of Organic Fuels on Flame Temperature of Magnesium/Oxidizer Mixtures[a]

Composition: Oxidizer 55% by weight; Magnesium 45% by weight; Shellac either 0 or 10% additional

	Approximate flame temperature, °C[b]	
	Oxidizer	
	$KClO_4$	$Ba(NO_3)_2$
Without shellac	3570	3510
With 10% shellac	2550	2550

[a]Reference 8.
[b]Temperature was measured 10 mm from the burning surface in the center of the flame.

Pyrotechnic flames typically fall in the 2000-3000°C range. Table 5.10 lists approximate values for some common classes of high-energy reactions [1].

TABLE 5.10 Maximum Flame Temperatures of Various Classes of Pyrotechnic Mixtures[a]

Type of composition	Maximum flame temperature, °C
Photoflash, illuminating	2500-3500
Solid rocket fuel	2000-2900
Colored flame mixtures	1200-2000
Smoke mixtures	400-1200

[a]Reference 1.

TABLE 5.11 Flame Temperatures for Oxidizer/Shellac Mixtures

Composition:	Oxidizer	75%
	Shellac	15%
	Sodium oxalate	10%

Oxidizer	Approximate flame temperature, °C[a]
Potassium chlorate, $KClO_3$	2160
Potassium perchlorate, $KClO_4$	2200
Ammonium perchlorate, NH_4ClO_4	2200
Potassium nitrate, KNO_3	1680

[a]Reference 8.

Binary mixtures of oxidizer with metallic fuel yield the highest flame temperatures, and the choice of oxidizer does not appear to make a substantial difference in the temperature attained. For compositions without metal fuels, this does not seem to be the case. Shimizu has collected data for a variety of compositions and has observed that potassium nitrate mixtures attain substantially lower flame temperatures than similar mixtures made with chlorate or perchlorate oxidizers. This result can be attributed to the substantially-endothermic decomposition of KNO_3 relative to the other oxidizers. Table 5.11 presents some of the Shimizu data [8].

A final factor that can limit the temperature of pyrotechnic flames is unanticipated high-temperature chemistry. Certain reactions that do not occur to any measurable extent at room temperature become quite probable at higher temperatures. An example of this is the reaction between carbon (C) and magnesium oxide (MgO). Carbon can be produced from organic molecules in the flame.

$$\underset{\text{(solid)}}{C} + \underset{\text{(solid)}}{MgO} \rightarrow \underset{\text{(gas)}}{CO} + \underset{\text{(gas above 1100°C)}}{Mg}$$

TABLE 5.12 Propagation Index Values for Pyrotechnic Mixtures[a]

Composition	% by weight	Heat of reaction, cal/gram	Ignition temperature, °C	Propagation index (cal/g-°C)
I. Boron igniter		1600	565	2.8
Boron	23.7			
Potassium nitrate	70.7			
Laminac resin	5.6			
II. Black powder		660[b]	330	2.0
Potassium nitrate	75			
Charcoal	15			
Sulfur	10			
III. Titanium igniter		740	520	1.4
Titanium	26			
Barium chromate	64			
Potassium perchlorate	10			
IV. Manganese delay		254	421	0.60
Manganese	41			
Lead chromate	49			
Barium chromate	10			

[a]Reference 9.
[b]Reference 1.

This is a strongly endothermic process, but it becomes possible at high temperature due to a favorable entropy change – formation of the random vapor state from solid reactants. Such reactions provide another reason for the lower flame temperatures achieved when organic binders are added to oxidizer/metal mixtures [3].

Propagation Index

A simple method for assessing the ability of a particular composition to burn is the "Propagation Index," originally proposed by McLain and later modified by Rose [3, 9]. The original McLain expression was

$$P_I = \frac{\Delta H_{reaction}}{T_{ignition}}$$

where P_I -- the Propagation Index -- is a measure of a mixture's tendency to sustain burning upon initial ignition by external stimulus. The equation contains the two main factors that determine burning ability – the amount of heat released by the chemical reaction (ΔH) and the ignition temperature of the mixture. If a substantial quantity of heat is released and the ignition temperature is low, then reignition from layer to layer should occur readily and propagation is likely. Conversely, mixtures with low heat output and high ignition temperature should propagate poorly, if at all. Propagation Index values for a variety of compositions are given in Table 5.12.

Rose recommended modifying the original McLain expression by the addition of terms for the pressed density of the composition and for the burning rate of the mixture. He reasoned, especially for delay compositions compressed in a tube, that ability to propagate should increase with increasing density, due to better heat transfer between grains of composition. Burning rate should also be a factor, he argued, because faster-burning mixtures should lose less heat to the surroundings than slower compositions [9].

REFERENCES

1. A. A. Shidlovskiy, *Principles of Pyrotechnics*, 3rd Ed., Moscow, 1964. (Translated by Foreign Technology Division, Wright-Patterson Air Force Base, Ohio, 1974.)

2. T. J. Barton, et al., "Factors Affecting the Ignition Temperature of Pyrotechnics," *Proceedings, Eighth International Pyrotechnics Seminar*, IIT Research Institute, Steamboat Springs, Colorado, July, 1982, p. 99.
3. J. H. McLain, *Pyrotechnics from the Viewpoint of Solid State Chemistry*, The Franklin Institute Press, Philadelphia, Penna., 1980.
4. H. Ellern, *Military and Civilian Pyrotechnics*, Chemical Publ. Co., Inc., New York, 1968.
5. U.S. Army Material Command, Engineering Design Handbook, Military Pyrotechnic Series, Part One, "Theory and Application," Washington, D.C., 1967 (AMC Pamphlet 706-185).
6. H. Henkin and R. McGill, *Ind. and Eng. Chem.*, *44*, 1391 (1952).
7. J. E. Tanner, "Effect of Binder Oxygen Content on Adiabatic Flame Temperature of Pyrotechnic Flares," RDTR No. 181, Naval Ammunition Depot, Crane, Indiana, August, 1972.
8. T. Shimizu, *Fireworks — The Art, Science and Technique*, pub. by T. Shimizu, distrib. by Maruzen Co., Ltd., Tokyo, 1981.
9. J. E. Rose, "Flame Propagation Parameters of Pyrotechnic Delay and Ignition Compositions," Report IHMR 71-168, Naval Ordnance Station, Indian Head, Maryland, 1971.
10. F. L. McIntyre, "A Compilation of Hazard and Test Data for Pyrotechnic Compositions," Report ARLCD-CR-80047, U.S. Army Armament Research and Development Command, Dover, NJ, 1980.

A "set piece" outlines the seal of the United States. The pyrotechnician creates pictures and messages by attaching hundreds of cigar-sized tubes, loaded with color-producing composition, to a wooden lattice secured in the ground. The pattern of the tubes and the choice of colors determine the picture that is produced. Fast-burning fuse—"quickmatch"—connects the tubes and permits rapid ignition of the entire pattern. Thread impregnated with fine black powder is covered by a loose-fitting paper wrapper to make quickmatch. The hot gas and flame is confined inside the paper sheath, and burning is very rapid. (Zambelli Internationale)

6
HEAT AND DELAY COMPOSITIONS

HEAT PRODUCTION

All pyrotechnic compositions evolve heat upon ignition, and this release of energy can be used to produce color, motion, smoke, and noise. There are applications as well for the chemically-produced heat itself, and these will be addressed in this chapter.

The use of incendiary mixtures in warfare can be traced back to ancient times, when it provided an effective means of assaulting well-fortified castles. Naval warfare was revolutionized by the use of flaming missiles to attack wooden ships, and much effort was put into improving the heat output, portability, and accuracy of these thermal weapons.

As both weaponry and the use of explosives for blasting developed, the need for a safe, reliable way to ignite these devices became obvious, and the concept of a pyrotechnic "delay" emerged. A variety of terms are used for materials that either ignite or provide a delay period between ignition of a device and the production of the main explosive or pyrotechnic effect. These include:

1. *Fuse*: A train of slow-burning powder (usually black powder), often covered with twine or twisted paper. Fuses are lit by a safety match or other hot object, and provide a time delay to permit the person igniting the device to retreat to a safe distance.

TABLE 6.1 Electric Match (Squib) Compositions[a]

Component	Formula	% by weight
1. Potassium chlorate	$KClO_3$	8.5
Lead mononitroresorcinate	$PbC_6H_3NO_4$	76.5
Nitrocellulose	-	15
2. Potassium chlorate	$KClO_3$	55
Lead thiocyanate	$Pb(SCN)_2$	45
3. Potassium perchlorate	$KClO_4$	66.6
Titanium	Ti	33.3

[a]Reference 1.

2. *Electric Match (Squib)*: A metal wire is coated with a dab of heat-sensitive composition. An electric current is passed through the wire, and the heat that is produced ignites the match composition. A burst of flame occurs that ignites a section of fuse or a charge of pyrotechnic composition. Squib compositions usually contain potassium chlorate (low ignition temperatures!). Lead mononitroresorcinate (LMNR) is also included in many squib mixtures. Several squib formulas are listed in Table 6.1.
3. *First Fire*: An easily-ignited composition is placed in limited quantity on top of the main pyrotechnic mixture. The first fire is reliably ignited by a fuse or squib, and the flame and hot residue that is produced then ignites the main charge. Black powder moistened with water containing a binder such as dextrine is used in the fireworks industry as a first fire, and also secures the fuse to the item. First fires are often referred to as "primes" – a term similar to another with a distinct meaning (see #5, below).
4. *Delay Composition*: A general term for a mixture that burns at a selected, reproducible rate, providing a time delay between activation and production of the main effect. A fuse containing a core of black powder is an example of a delay. Highly-reproducible delay mixtures are needed for military applications, and much research effort has been put into developing reliable compositions.

TABLE 6.2 Typical Primer Mixtures[a]

Component	Formula	% by weight	Note
1. Potassium chlorate	$KClO_3$	45	Stab primer
Lead thiocyanate	$Pb(SCN)_2$	33	
Antimony sulfide	Sb_2S_3	22	
2. Potassium chlorate	$KClO_3$	33	Stab primer
Antimony sulfide	Sb_2S_3	33	
Lead azide	$Pb(N_3)_2$	29	
Carborundum	-	5	
3. Potassium chlorate	$KClO_3$	50	Percussion primer
Lead peroxide	PbO_2	25	
Antimony sulfide	Sb_2S_3	20	
Trinitrotoluene	$C_7H_5N_3O_6$	5	
4. Potassium chlorate	$KClO_3$	50	Percussion primer
Zirconium	Zr	50	

[a]Reference 1.

5. *Primer*: A term for the device used to ignite smokeless powder in small arms ammunition. An impact-sensitive composition is used. When struck by a metal firing pin, a primer emits a burst of flame capable of igniting the propellant charge. Several typical primer mixtures are given in Table 6.2.
6. *Friction Igniter*: A truly "self-contained" device should be ignitible without the need for a safety match or other type of external ignition source. Highway flares (fusees), other types of distress signals, and some military devices use a friction ignition system. The fusee uses a two-part igniter; when the two surfaces are rubbed together, a flame is produced and the main composition is ignited. Typically, the scratcher portion of these devices contains red phosphorus and the matchhead mixture contains potassium chlorate ($KClO_3$) and a good fuel. Several friction igniter systems are given in Table 6.3.

TABLE 6.3 Friction Igniter Mixtures

Component	Formula	% by weight	Reference
1. Main composition			
Potassium chlorate	$KClO_3$	60	1
Antimony sulfide	Sb_2S_3	30	
Resin	-	10	
Striker			
Red phosphorus	P	56	
Ground glass	SiO_2	24	
Phenol/formaldehyde resin	$(C_{13}H_{12}O_2)_7$	20	
2. Main composition			
Shellac	-	40	2
Strontium nitrate	$Sr(NO_3)_2$	3	
Quartz	SiO_2	6	
Charcoal	C	2	
Potassium perchlorate	$KClO_4$	14	
Potassium chlorate	$KClO_3$	28	
Wood flour	-	5	
Marble dust	$CaCO_3$	2	
Striker			
Lacquer	-	61	
Pumice	-	2.2	
Red phosphorus	P	26	
Butyl acetate	$C_6H_{12}O_2$	10.8	

DELAY COMPOSITIONS

The purpose of a delay composition is obvious – to provide a time delay between ignition and the delivery of the main effect. Crude delays can be made from loose powder, but a compressed column is capable of much more reproducible performance. The burning rates of delay mixtures range from very fast (millimeters/millisecond) to slow (millimeters/second).

Black powder was the sole delay mixture available for several centuries. The development and use of "safety fuse" containing a black powder core significantly improved the safety record of the blasting industry. However, the development of modern, long-range, high-altitude projectiles created a requirement for a new generation of delay mixtures. Black powder, under specified conditions, gives reproducible burning rates at ground level. However, it produces a considerable quantity of gas upon ignition (approximately 50% of the reaction products are gaseous), and its burning rate will therefore show a significant dependence on external pressure (faster burning as external pressure increases). To overcome this pressure dependence, researchers set out to develop "gasless" delays – mixtures that evolve heat and burn at reproducible rates with the formation of *only* solid and liquid products. Such mixtures show little, if any, variation of burning rate with pressure.

One could begin such a project by setting down the requirements for an "ideal" delay mixture [4]:

1. The mixture should be stable during preparation and storage. Materials of low hygroscopicity must be used.
2. The mixture should be readily ignitible from a modest ignition stimulus.
3. There must be minimum variation in the burning rate of the composition with changes in external temperature and pressure. The mixture must readily ignite and reliably burn at low temperature and pressure.
4. There should be a minimum change in the burning rate with small percentage changes in the various ingredients.
5. There must be reproducible burning rates, both within a batch and between batches.

The newer "gasless" delays are usually a combination of a metal oxide or chromate with an elemental fuel. The fuels are metals or high-heat nonmetallic elements such as silicon or boron. If an organic binder (e.g., nitrocellulose) is used, the resulting mixture will be "low gas" rather than "gasless," due to the carbon dioxide (CO_2), carbon monoxide (CO), and nitrogen (N_2) that will form upon combustion of the binder. If a truly "gasless" mixture is required, leave out all organic materials!

If a fast burning rate is desired, a metallic fuel with high heat output per gram should be selected, together with an oxidizer of low decomposition temperature. The oxidizer should also have a small endothermic – or even better, exothermic – heat of decomposition. For slower delay mixtures, metals with less heat output per gram should be selected, and oxidizers with higher

TABLE 6.4 Typical Delay Compositions[a]

Component	Formula	% by weight	Burning rate, cm/second
1. Red lead oxide	Pb_3O_4	85	1.7 (10.6 ml/g of gas)
Silicon	Si	15	
Nitrocellulose/acetone	-	1.8	
2. Barium chromate	$BaCrO_4$	90	5.1 (3.1 ml/g of gas)
Boron	B	10	
3. Barium chromate	$BaCrO_4$	40	– (4.3 ml/g of gas)
Potassium perchlorate	$KClO_4$	10	
Tungsten	W	50	
4. Lead chromate	$PbCrO_4$	37	0.30 (18.3 ml/g of gas)
Barium chromate	$BaCrO_4$	30	
Manganese	Mn	33	
5. Barium chromate	$BaCrO_4$	80	0.16 (0.7 ml/g of gas)
Zirconium-nickel alloy (50/50)	Zr-Ni	17	
Potassium perchlorate	$KClO_4$	3	

[a]Reference 1.

decomposition temperatures and more endothermic heats of decomposition should be chosen. By varying the oxidizer and fuel, it is possible to create delay compositions with a wide range of burning rates. Table 6.4 lists some representative delay mixtures.

Using this approach, lead chromate (melting point 844°C) would be expected to produce faster burning mixtures than barium chromate (higher melting point), and barium peroxide (melting point 450°C) should react more quickly than iron oxide (Fe_2O_3, melting point 1565°C). Similarly, boron (heat of combustion = 14.0 kcal/gram) and aluminum (7.4 kcal/gram) should form quicker delay compositions than tungsten (1.1 kcal/gram) or iron (1.8 kcal/gram). Tables 3.2, 3.4, and 3.5 can be used to estimate

TABLE 6.5 The Barium Chromate/Boron System – Effect of % Boron on Burning Time[a]

% B	Average burning time seconds/gram	Heat of reaction cal/gram
3	3.55	278
5	.51	420
7	.33	453
10	.24	515
13	.21	556
15	.20	551
17	.21	543
21	.22	526
25	.27	497
30	.36	473
35	.64	446
40	1.53	399
45	3.86	364

[a]Reference 4.

the relative burning rates of various delay candidates. For high reactivity, look for low melting point, exothermic or small endothermic heat of decomposition (in the oxidizer), and high heat of combustion (in the fuel).

The ratio of oxidizer to fuel can be altered for a given binary mixture to achieve substantial changes in the rate of burning. The fastest burning rate should correspond to an oxidizer/fuel ratio near the stoichiometric point, with neither component present in substantial excess. Data have been published for the barium chromate/boron system. Table 6.5 gives the burn time and heat output per gram for this system [4].

McLain has proposed that the maximum in performance centered at approximately 15% boron by weight indicates that the principal pyrotechnic reaction for the $BaCrO_4/B$ system is

$$4\ B + BaCrO_4 \rightarrow 4\ BO + Ba + Cr$$

Although B_2O_3 is the expected oxidation product from boron in a room temperature situation, the lower oxide, BO, appears to

TABLE 6.6 A Ternary Delay Mixture – The $PbCrO_4$/$BaCrO_4$/Mn System[a]

Mixture	% Manganese, Mn	% Lead chromate	% Barium chromate	Burning rate, cm/second
I.	44	53	3	0.69
II.	39	47	14	0.44
III.	37	43	20	0.29
IV.	33	36	31	0.19

[a]Reference 2. Data from H. Ellern, *Military and Civilian Pyrotechnics*, Chemical Publishing Co., Inc., New York, 1968.

be more stable at the high reaction temperature of the burning delay mixture [2].

A small percentage of fuel in excess of the stoichiometric amount increases the burning rate for most delay mixtures, presumably through increased thermal conductivity for the composition. The propagation of burning is enhanced by the additional metal, especially in the absence of substantial quantities of hot gas to aid in the propagation of burning. Air oxidation of the excess metal fuel can also contribute additional heat to increase the reaction rate *if* the burning composition is exposed to the atmosphere.

The rate of burning of ternary mixtures can similarly be affected by varying the percentages of the components. Table 6.6 presents data for a three-component delay composition. In this study, a decrease in the burning rate (in cm/second) is observed as the metal percentage is lowered (giving poorer thermal conductivity) and the percentage of higher-melting oxidizer ($BaCrO_4$) is increased at the expense of the lower-melting, more reactive lead chromate, $PbCrO_4$.

Table 6.7 illustrates this same concept for the molybdenum/barium chromate/potassium perchlorate system. Here, $KClO_4$ is the better oxidizer.

Contrary to the behavior expected for "gassy" mixtures, the rate of burning for gasless compositions is expected to *increase* (in units of grams reacting per second) as the consolidation pressure is increased. "Gasless" delays propagate via heat transfer

TABLE 6.7 The $BaCrO_4/KClO_4/Mo$ System[a]

Mixture	% Barium chromate, $BaCrO_4$	% Potassium perchlorate, $KClO_4$	% Molybdenum, Mo	Burning rate, cm/second
I.	10	10	80	25.4
II.	40	5	55	1.3
III.	55	10	35	0.42
IV.	65	5	30	0.14

[a]Reference 2. Data from H. Ellern, *Military and Civilian Pyrotechnics*, Chemical Publishing Co., Inc., New York, 1968.

down the column of pyrotechnic material, and the thermal conductivity of the mixture plays a significant role. As the density of the mixture increases due to increased loading pressure, the components are pressed closer together and better heat transfer occurs. Table 4.6 presented data for the barium chromate/boron system, showing the modest increase that occurs as the loading pressure is raised.

IGNITION COMPOSITIONS AND FIRST FIRES

Compositions with high ignition temperatures (i.e., above 600°C) can be difficult to ignite using solely the "spit" from a black powder fuse or similar mild ignition stimulus. In such situations, an initial charge of a more-readily-ignitible material, called a "first fire," is frequently used. The requirements for such a mixture include [3]:

1. Reliable ignitibility from a small thermal impulse such as a fuse. The ignition temperature of a "first fire" should be 500°C or less.
2. The mixture should attain a high reaction temperature, well above the ignition temperature of the main composition. Metal fuels are usually used when *high* reaction temperatures are needed.

3. A mixture that forms a hot, liquid slag is preferred. Such slag will provide considerable surface contact with the main composition, facilitating ignition. The production of hot gas will usually produce good ignition behavior on the ground, but reliability will deteriorate at higher altitudes. Liquid and solid products provide better heat retention to aid ignition under these conditions.
4. A slower-burning mixture is preferred over a more rapid one. The slower release of energy allows for better heat transfer to the main composition. Also, most "first fires" are pressed into place or added as moist pastes (that harden on drying), rather than used as faster-burning loose powders.

Potassium nitrate is frequently used in igniters and first fires. Compositions made with this oxidizer tend to have low ignition temperatures (typically below 500°C), and yet the mixtures are reasonably safe to prepare, use in production, and store. Potassium chlorate formulations also tend to have low ignition temperatures, but they are considerably more sensitive (and hazardous).

Potassium nitrate mixed with charcoal can be used for ignition, as can black powder worked into a paste with water and a little dextrine. Shidlovskiy reports that the composition

KNO_3, 75
Mg, 15
Iditol, 10 (iditol is a phenol/formaldehyde resin)

works well as an igniter mixture [3]; the solid magnesium oxide (MgO) residue aids in igniting the main composition. Boron mixed with potassium nitrate is a frequently-used, effective igniter mixture, as is the combination of iron oxide with zirconium metal and diatomaceous earth (commonly known as A-1A ignition mixture). Table 6.8 lists a variety of formulations that have been published.

THERMITE MIXTURES

Thermites are mixtures that produce a high heat concentration, usually in the form of molten products. Thermite compositions contain a metal oxide as the oxidizer and a metal -- usually aluminum -- as the fuel, although other active metals may be used.

TABLE 6.8 Ignition and First Fire Compositions

Mixture	Component	Formula	% by weight	Note
I.	Barium peroxide	BaO_2	80	Reference 3
	Magnesium	Mg	18	Solid BaO particles aid in ignition
	Binder	-	2	
II.	Iron oxide	Fe_2O_3	65	A-1A mixture, a "gasless" igniter
	Zirconium	Zr	25	
	Diatomaceous earth	-	10	
III.	Black powder	$KNO_3/S/C$	75	Reference 3
	Potassium nitrate	KNO_3	12	
	Zirconium	Zr	13	
IV.	Potassium nitrate	KNO_3	71	Reference 3
	Boron	B	24	
	Rubber	-	5	
V.	Red lead oxide	Pb_3O_4	50	Reference 1
	Titanium	Ti	25	
	Silicon	Si	25	
VI.	Sodium nitrate	$NaNO_3$	47	Reference 1
	Sugar	$(C_{12}H_{22}O_{11})_n$	47	
	Charcoal	C	6	
VII.	Barium peroxide	BaO_2	88	Reference 3
	Magnesium	Mg	12	Thermite igniter

A minimum amount of gas is produced, enabling the heat of reaction to concentrate in the solid and liquid products. High reaction temperatures can be achieved in the absence of volatile materials; typically, values of 2000-2800°C are reached [3]. A metal product such as iron, with a wide liquid range (melting point 1535°C, boiling point 2800°C) produces the best thermite behavior. Upon ignition, a thermite mixture will form aluminum oxide and the metal corresponding to the starting metal oxide:

$$Fe_2O_3 + 2\ Al \rightarrow Al_2O_3 + 2\ Fe$$

Thermite mixtures have found application as incendiary compositions and spot-welding mixtures. They are also used for the intentional demolition of machinery and for the destruction of documents. Thermites are usually produced without a binder (or with a minimum of binder), because the gaseous products resulting from the combustion of the organic binder will carry away heat and cool the reaction.

Iron oxide (Fe_2O_3 or Fe_3O_4) with aluminum metal is the classic thermite mixture. The particle size of the aluminum should be somewhat coarse to prevent the reaction from being too rapid. Thermites tend to be quite safe to manufacture, and they are rather insensitive to most ignition stimuli. In fact, the major problem with most thermites is *getting* them to ignite, and a strong first fire is usually needed.

Calorific data for a variety of aluminum thermite mixtures are given in Table 6.9.

PROPELLANTS

The production of hot gas to lift and move objects, using a pyrotechnic system, began with the development of black powder. Rockets were in use in Italy in the 14th century [5], and cannons were developed at about the same time. The development of aerial fireworks was a logical extension of cannon technology.

Black powder remained the sole propellant available for military and civilian applications until well into the 19th century. A number of problems associated with the use of black powder stimulated efforts to locate replacements:

1. Substantial variation in burning behavior from batch to batch. The better black powder factories produced good powder *if* they paid close attention to the purity of their

TABLE 6.9 Calorific Data for Thermite Mixtures[a]

Oxidizers	Formula	% Active oxygen by weight	% Al by weight in thermite mixture	$\Delta H_{reaction}$, kcal/gram
Silicon dioxide	SiO_2	53	37	.56
Chromium(III) oxide	Cr_2O_3	32	26	.60
Manganese dioxide	MnO_2	37	29	1.12
Iron oxide	Fe_2O_3	30	25	.93
Iron oxide	Fe_3O_4	28	24	.85
Cupric oxide	CuO	20	19	.94
Lead oxide (red)	Pb_3O_4	9	10	.47

[a]Reference 3.

starting materials, used one source of charcoal, and did not vary the extent of mixing or the amount of moisture contained in their product.

2. Black powder has a relatively low gas output. Only about 50% of the products are gaseous; the remainder are solids.
3. The solid residue from black powder is highly alkaline (strongly basic), and it is quite corrosive to many materials.

"Pyrodex" is a patented pyrotechnic composition designed to fulfill many of the functions of black powder. It contains the three ingredients found in black powder plus binders and burning rate modifiers that make the material somewhat less sensitive and slower burning. A greater degree of confinement is required to obtain performance comparable to "normal" black powder [6].

The advantages of black powder and Pyrodex include good ignitibility, moderate cost, ready availability of the ingredients, and a wide range of uses (fuse powder, delay mixture, propellant, and explosive) depending on the degree of confinement.

As propellant technology developed, the ideal features for a better material became evident:

1. A propellant that can safely be prepared from readily-available materials at moderate cost.
2. A material that readily ignites, but yet is stable during storage.
3. A mixture that forms the maximum quantity of low molecular weight gases upon burning, with minimum solid residue.
4. A mixture that reacts at the highest possible temperature, to provide maximum thrust.

The late 19th century saw the development of a new family of "smokeless" powders, as modern organic chemistry blossomed and the nitration reaction became commercially feasible. Two "esters" – nitrocellulose and nitroglycerine – became the major components of these new propellants. An ester is a compound formed from the reaction between an acid and an alcohol. Figure 6.1 illustrates the formation of NC and NG from nitric acid and the precursor alcohols cellulose and glycerine.

"Single base" smokeless powder, developed mainly in the United States, uses only nitrocellulose. "Double base" smokeless powder, developed in Europe, is a blend of nitrocellulose and nitroglycerine. "Cordite," a British development, consists of 65% NC, 30% NG, and 5% mineral jelly. The mineral jelly (a hydrocarbon material) functions as a coolant and produces substantial amounts of CO_2, CO, and H_2O gas to improve the propellant characteristics. "Triple base" smokeless powder, containing nitroguanidine as a third component with nitroglycerine and nitrocellulose is also manufactured.

An advantage of the smokeless powders is their ability to be *extruded* during the manufacturing process. Perforated grains can be produced that simultaneously burn inwardly and outwardly such that a constant burning surface area and constant gas production are achieved.

Nitrocellulose does not contain sufficient internal oxygen for complete combustion to CO_2, H_2O, and N_2, while nitroglycerine contains excess oxygen [7]. The double base smokeless propellants therefore achieve a slightly more complete combustion and benefit from the substantial exothermicity of NG (1486 calories/gram) [7].

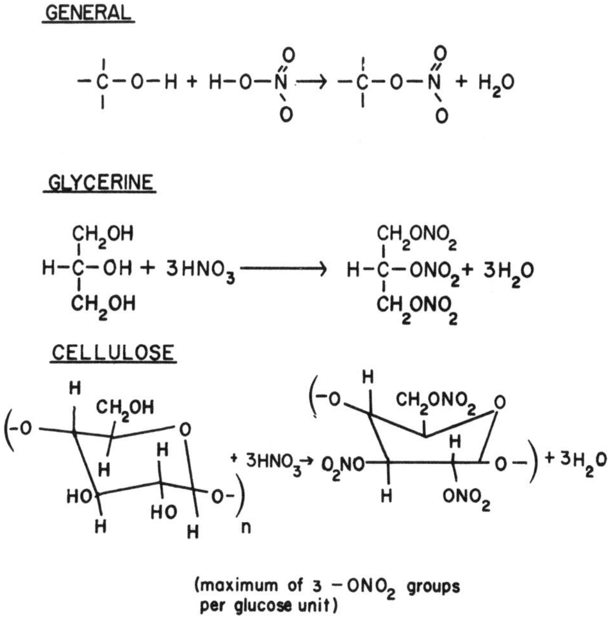

GENERAL

$$-\overset{|}{\underset{|}{C}}-O-H + H-O-N\overset{O}{\underset{O}{\lessgtr}} \longrightarrow -\overset{|}{\underset{|}{C}}-O-N\overset{O}{\underset{O}{\lessgtr}} + H_2O$$

GLYCERINE

$$CH_2OH-CH(OH)-CH_2OH + 3HNO_3 \longrightarrow CH_2ONO_2-CH(ONO_2)-CH_2ONO_2 + 3H_2O$$

CELLULOSE

$$(\text{cellulose})_n + 3HNO_3 \rightarrow (\text{cellulose nitrate})_n + 3H_2O$$

(maximum of 3 $-ONO_2$ groups per glucose unit)

FIG. 6.1 The nitration reaction. Organic compounds containing the $-OH$ functional group are termed "alcohols." These compounds react with nitric acid to produce a class of compounds known as "nitrate esters." Nitroglycerine and nitrocellulose are among the numerous explosive materials produced using this reaction.

Smokeless powders are widely used today as the propellants for small arms ammunition as well as for artillery projectiles. Black powder remains the propellant used by the fireworks industry for sky rockets and for aerial shells fired from mortars, and is used in a variety of military applications as well.

The larger rockets used by the military and in the space program require *tremendous* thrust to lift off successfully, and combinations of liquid fuel engines (e.g., liquid hydrogen peroxide (H_2O_2) and hydrazine (N_2H_4)) with solid propellant pyrotechnic boosters are used to lift enormous vehicles such as the Space

Shuttle. The pyrotechnic boosters used for these launches typically contain:

1. *A solid oxidizer*: Ammonium perchlorate (NH_4ClO_4) is the current favorite due to the high percentage of gaseous products it forms upon reaction with a fuel.
2. *A small percentage of light, high-energy metal*: This metal produces solid combustion products that do not aid in achieving thrust, but the considerable heat evolved by the burning of the metal raises the temperature of the other gaseous products. Aluminum and magnesium are the metals most commonly used.
3. *An organic fuel that also serves as binder and gas-former*: Liquids that polymerize into solid masses are preferred, for simpler processing, and a binder with low oxygen content is desirable to maximize heat production.

A negative oxygen balance is frequently designed into these propellant mixtures to obtain CO gas in place of CO_2. CO is lighter and will produce greater thrust, all other things being equal. However, the full oxidation of carbon atoms to CO_2 evolves more heat, so some trial-and-error is needed to find the optimum ratio of oxidizer and fuel [8].

Propellant compositions are also used in numerous "gas generator" devices, where the production of gas pressure is used to drive pistons, trigger switches, eject pilots from aircraft, and perform an assortment of other critical functions. The military and the aerospace industry use many of these items, which can be designed to function rapidly and can be initiated remotely.

REFERENCES

1. F. L. McIntyre, A Compilation of Hazard and Test Data for Pyrotechnic Compositions," Report ARLCD-CR-80047, U.S. Army Armament Research and Development Command, Dover, NJ, 1980.
2. J. H. McLain, *Pyrotechnics from the Viewpoint of Solid State Chemistry*, The Franklin Institute Press, Philadelphia, Penna., 1980.
3. A. A. Shidlovskiy, *Principles of Pyrotechnics*, 3rd Ed., Moscow, 1964. (Translated by Foreign Technology Division, Wright-Patterson Air Force Base, Ohio, 1974.)

4. U.S. Army Material Command, Engineering Design Handbook, Military Pyrotechnic Series, Part One, "Theory and Application," Washington, D.C., 1967 (AMC Pamphlet 706-185).
5. J. R. Partington, *A History of Greek Fire and Gunpowder*, W. Heffer & Sons, Ltd., Cambridge, England, 1960.
6. G. D. Barrett, "Venting of Pyrotechnics Processing Equipment," *Proceedings, Explosives and Pyrotechnics Applications Section*, American Defense Preparedness Assn., Los Alamos, New Mexico, October, 1984.
7. "Military Explosives," U.S. Army and U.S. Air Force Technical Manual TM 9-1300-214, Washington, D.C., 1967.
8. R. F. Gould (Ed.), *Advanced Propellant Chemistry*, American Chemical Society Publications, Washington, D.C., 1966.

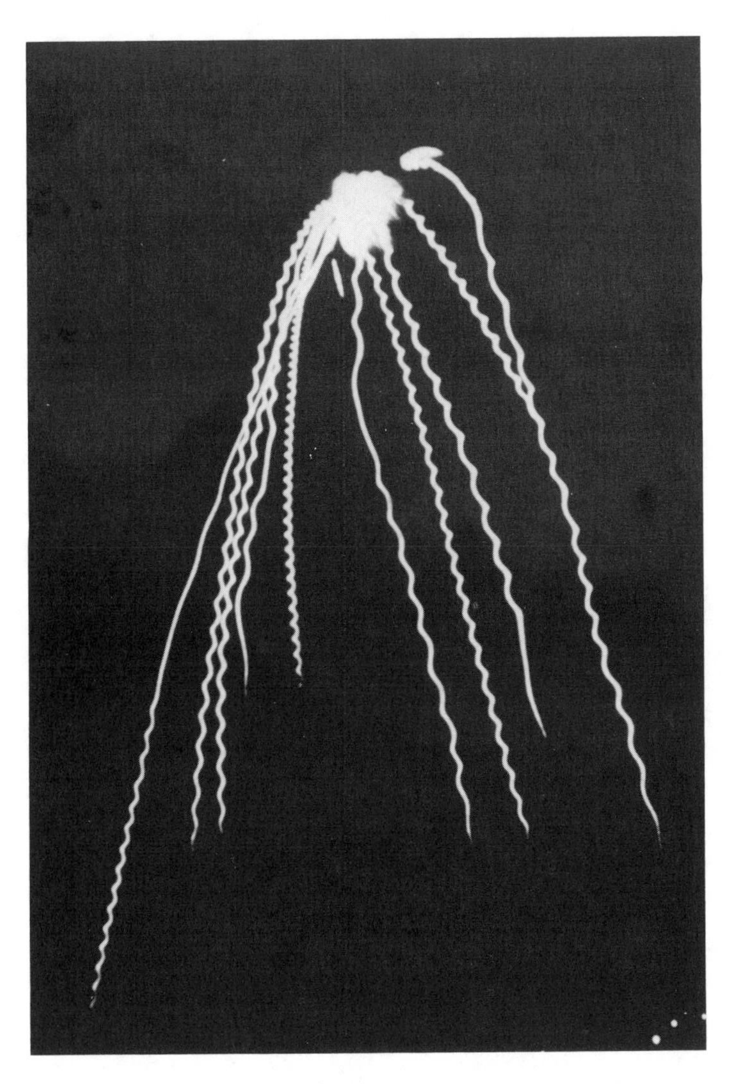

A "weeping willow" aerial shell bursts high in the sky and leaves its characteristic pattern as the large, slow-burning stars descend to the ground. Charcoal is frequently used to produce the attractive gold color, with potassium nitrate selected as the oxidizer to achieve a slow-burning mixture. (Zambelli Internationale)

7

COLOR AND LIGHT PRODUCTION

The production of bright light and vivid color is the primary purpose of many pyrotechnic compositions. Light emission has a variety of applications, ranging from military signals and highway distress flares to spectacular aerial fireworks. The basic theory of light emission was discussed in Chapter 2, and several good articles have been published dealing with the chemistry and physics of colored flames [1, 2].

The quantitative measurement of light intensity (candle power) at any instant and the light integral (total energy emitted, with units of candle-seconds/gram) can be affected by a variety of test parameters such as container diameter, burning rate, and the measuring equipment. Therefore, comparisons between data obtained from different reports should be viewed with caution.

WHITE LIGHT COMPOSITIONS

For white-light emission, a mixture is required that burns at high temperature, creating a substantial quantity of excited atoms or molecules in the vapor state together with incandescent solid or liquid particles. Incandescent particles emit a broad range of wavelengths in the visible region of the electromagnetic spectrum, and white light is perceived by the viewer. Intense emission from sodium atoms in the vapor state, excited to higher-energy electronic states by high flame temperature, is the principal light source in the sodium nitrate/magnesium/organic binder flare compositions widely used by the military [3, 4].

Magnesium or aluminum fuels are found in most white-light compositions. These metals evolve substantial heat upon oxidation, and the high-melting magnesium oxide (MgO) and aluminum oxide (Al_2O_3) reaction products are good light emitters at the high reaction temperatures that can be achieved using these fuels. Titanium and zirconium metals are also good fuels for white-light compositions.

In selecting an oxidizer and fuel for a white-light mixture, a main consideration is maximizing the heat output. A value of 1.5 kcal/gram has been given by Shidlovskiy as the minimum for a usable illuminating composition [5]. A flame temperature of less than 2000°C will produce a minimum amount of white light by emission from incandescent particles or from excited gaseous sodium atoms.

Therefore, the initial choice for an oxidizer is one with an *exothermic* heat of decomposition such as potassium chlorate ($KClO_3$). However, mixtures of both chlorate and perchlorate salts with active metal fuels are *too* ignition-sensitive for commercial use, and the less-reactive – but safer – nitrate compounds are usually selected. Potassium perchlorate is used with aluminum and magnesium in some "photoflash" mixtures; these are extremely reactive compositions, with velocities in the explosive range.

The nitrates are considerably *endothermic* in their decomposition and therefore deliver less heat than chlorates or perchlorates, but they can be used with less fear of accidental ignition. Barium nitrate is often selected for white-light mixtures. The barium oxide (BaO) product formed upon reaction is a good, broad-range molecular emitter in the vapor phase (the boiling point of BaO is ca. 2000°C), and condensed particles of BaO found in the cooler parts of the flame are also good emitters of incandescent light.

Sodium nitrate is another frequent choice. It is quite hygroscopic however, so precautions must be taken during production and storage to exclude moisture. Sodium nitrate produces good heat output per gram due to the low atomic weight (i.e., 23) of sodium, and the intense flame emission from atomic sodium in the vapor state contributes substantially to the total light intensity. Potassium nitrate, on the other hand, is not a good source of atomic or molecular emission, and it is rarely – if ever – used as the sole oxidizer in white-light compositions.

Magnesium metal is the fuel found in most military illuminating compositions, as well as in many fireworks devices. Aluminum and titanium metals, the magnesium/aluminum alloy "magnalium," and antimony sulfide (Sb_2S_3) are used for white light effects in many fireworks mixtures. Several published formulas for white light compositions are given in Table 7.1.

The ratio of ingredients, as expected, will affect the performance of the composition. Optimum performance is anticipated near the stoichiometric point, but an excess of metallic fuel usually increases the burning rate and light emission intensity. The additional metal increases the thermal conductivity of the mixture, thereby aiding burning, and the excess fuel – especially a volatile metal such as magnesium (boiling point 1107°C) – can vaporize and burn with oxygen in the surrounding air to produce extra heat and light. The sodium nitrate/magnesium system is extensively used for military illuminating compositions. Data for this system are given in Table 7.2.

The anticipated reaction between sodium nitrate and magnesium is

	5 Mg +	2 $NaNO_3$ → 5 MgO + Na_2O + N_2
grams	121.5	170
% by weight	41.6	58.4 (for a stoichiometric mixture)

Formula A in Table 7.2 therefore contains an excess of oxidizer. It is the slowest burning mixture and produces the least heat. Formula B is very close to the stoichiometric point. Formula C contains excess magnesium and is the most reactive of the three; the burning of the excess magnesium in air must contribute substantially to the performance of this composition.

A significant altitude effect will be shown by these illuminating compositions, especially those containing excess metal. The decreased atmospheric pressure – and therefore less oxygen – at higher altitudes will slow the burning rate as the excess fuel will not be consumed as efficiently.

"Photoflash" Mixtures

To produce a burst of light of short duration, a composition is required that will react very rapidly. Fine particle sizes are used for the oxidizer and fuel to increase reactivity, *but* sensitivity is also enhanced at the same time. Therefore, these mixtures are quite hazardous to prepare, and mixing operations should always be carried out remotely. Several representative photoflash mixtures are given in Table 7.3.

An innovation in military photoflash technology was the development of devices containing fine metal powders *without* any oxidizer. A high-explosive bursting charge is used instead. This charge, upon ignition, scatters the metal particles at high

TABLE 7.1 White Light Compositions

	Oxidizer (% by weight)		Fuel (% by weight)		Other (% by weight)		Reference
I.	Barium nitrate, $Ba(NO_3)_2$	38.3	Magnesium, Mg	26.9	Wax	6.7	7
	Potassium nitrate, KNO_3	25.2			Oil	2.9	
II.	Sodium nitrate, $NaNO_3$	44	Magnesium	50	Laminac	6	8
III.	Teflon, $(-CF_2-CF_2-)_n$	46	Magnesium	54	Nitrocellulose	2.6	8
IV.	Sodium nitrate, $NaNO_3$	53	Aluminum	35	Vinyl alcohol/acetate resin ("VAAR")	5	8
			Tungsten, W	7			
V.	Potassium perchlorate, $KClO_4$	64	Antimony, Sb	13	Gum	10	6
	Potassium nitrate, KNO_3	13					
VI.	Potassium nitrate, KNO_3	65	Sulfur	20	Fine black powder	5	6
			Antimony	10			
VII.	Ammonium perchlorate, NH_4ClO_4	40	Antimony sulfide, Sb_2S_3	14	Wood meal	5	6
	Potassium perchlorate, $KClO_4$	30	Starch	11			

TABLE 7.2 The Sodium Nitrate/Magnesium System[a]

	% Sodium nitrate	% Magnesium	Linear burning rate, mm/sec	Heat of reaction, kcal/gram
A.	70	30	4.7	1.3
B.	60	40	11.0	2.0
C.	50	50	14.3	2.6

[a]Reference 5.

temperature and they are then air-oxidized to produce light emission. No hazardous mixing of oxidizer and fuel is required to prepare these illuminating devices.

SPARKS

The production of brilliant sparks is one of the principal effects available to the fireworks manufacturer and to the "special effects" industry. Sparks occur during the burning of many pyrotechnic compositions, and they may or may not be a desired feature.

Sparks are produced when liquid or solid particles – either original components of a mixture or particles created at the burning surface – are ejected from the composition by gas pressure produced during the high-energy reaction. These particles – heated to incandescent temperatures – leave the flame area and proceed to radiate light as they cool off or continue to react with atmospheric oxygen. The particle size of the fuel will largely determine the quantity and size of sparks; the larger the particle size, the larger the sparks are likely to be. A combination of fine fuel particles for heat production with larger particles for the spark effect is often used by manufacturers.

Metal particles – especially aluminum, titanium, and "magnalium" alloy – produce good sparks that are white in appearance. Charcoal of sufficiently large particle size also works well, producing sparks with a characteristic orange color. Sparks from iron particles vary from gold to white, depending on the

TABLE 7.3 Photoflash Mixtures

Oxidizer (% by weight)		Fuel (% by weight)		Reference
I. Potassium perchlorate, $KClO_4$	40	Magnesium Aluminum	34 26	7
II. Potassium perchlorate, $KClO_4$	40	Magnesium aluminum alloy, "Magnalium" (50/50)	60	7
III. Potassium perchlorate, $KClO_4$ Barium nitrate, $Ba(NO_3)_2$	30 30	Aluminum	40	7
IV. Barium nitrate, $Ba(NO_3)_2$	54.5	Magnalium Aluminum	45.5 4	8

reaction temperature; they are the brilliant sparks seen in the popular "gold sparkler" ignited by millions of people on the 4th of July.

Magnesium metal does not produce a good spark effect. The metal has a low boiling point (1107°C), and therefore tends to vaporize and completely react in the pyrotechnic flame [6]. "Magnalium" can produce good sparks that burn in air with a novel, crackling sound. Several spark-producing formulas are given in Table 7.4. Remember, the particle size of the fuel is very important in producing sparks – experimentation is needed to find the ideal size.

For a good spark effect, the fuel must contain particles large enough to escape from the flame prior to complete combustion. Also, the oxidizer must not be *too* effective, or complete reaction will occur in the flame. Charcoal sparks are difficult to achieve with the hotter oxidizers; potassium nitrate (KNO_3) – with its low flame temperatures – works best. Some gas production is required to achieve a good spark effect by assisting in the ejection of particles from the flame. Charcoal, other organic fuels and binders, and the nitrate ion can provide gas for this purpose.

TABLE 7.4 Spark-Producing Compositions

Composition	% by weight	Effect	Reference
I. Potassium nitrate, KNO_3	58	Gold sparks	6
Sulfur	7		
Pure charcoal	35		
II. Barium nitrate, $Ba(NO_3)_2$	50	Gold sparks (gold sparkler)	6
Steel filings	30		
Dextrine	10		
Aluminum powder	8		
Fine charcoal	0.5		
Boric acid	1.5		
III. Potassium perchlorate, $KClO_4$	42.1	White sparks	9
Titanium	42.1		
Dextrine	15.8		
(Make a paste from dextrine and water, then mix in oxidizer and fuel)			
IV. Potassium perchlorate, $KClO_4$	50	White sparks "waterfalls" effect	6
"Bright" aluminum powder	25		
"Flitter" aluminum, 30-80 mesh	12.5		
"Flitter" aluminum, 5-30 mesh	12.5		

Note: Particle size of the fuel is very important in determining the size of the sparks.

FLITTER AND GLITTER

Several interesting visual effects can be achieved by careful selection of the fuel and oxidizer for a spark-producing composition.

A thorough review article discussing this topic in detail – with numerous formulas – has been published [10].

"Flitter" refers to the large white sparks obtained from the burning of large aluminum flakes. These flakes burn continuously upon ejection from the flame, creating a beautiful white effect, and they are used in a variety of fireworks items.

"Glitter" is the term given to the effect produced by molten droplets which, upon ejection from the flame, ignite in air to produce a brilliant flash of light. A nitrate salt (KNO_3 is best) and sulfur or a sulfide compound appear to be essential for the glitter phenomenon to be achieved. It is likely that the low melting point (334°C) of potassium nitrate produces a liquid phase that is responsible, at least in part, for this effect. Several "glitter" formulas are given in Table 7.5. The ability of certain compositions containing magnesium or magnalium alloy to burn in a pulsing, "strobe light" manner is a novel phenomenon believed to involve two distinct reactions. A slow, "dark" process occurs until sufficient heat is generated to initiate a fast, light-emitting reaction. Dark and light reactions continue in an alternate manner, generating the strobe effect [11, 12].

COLOR

Introduction

Certain elements and compounds, when heated to high temperature, have the unique property of emitting lines or narrow bands of light in the visible region (380-780 nanometers) of the electromagnetic spectrum. This emission is perceived as color by an observer, and the production of colored light is one of the most important goals sought by the pyrotechnic chemist. Table 7.6 lists the colors associated with the various regions of the visible spectrum. The *complementary* colors – perceived if white light *minus* a particular portion of the visible spectrum is viewed – are also given in Table 7.6.

To produce color, heat (from the reaction between an oxidizer and a fuel) and a color-emitting species are required. Sodium compounds added to a heat mixture will impart a yellow color to the flame. Strontium salts will yield red, barium and copper compounds can give green, and certain copper-containing mixtures will produce blue. Color can be produced by emission of a narrow band of light (e.g., light in the range 435-480 nanometers is perceived as blue), or by the emission of several ranges of light that combine to yield a particular color. For example, the mixing of

TABLE 7.5 Glitter Formulas[a]

Component	% by weight	Effect	Note
I. Potassium nitrate, KNO_3	55	Good white glitter	Used in aerial stars
"Bright" aluminum powder	5		
Dextrine	4		
Antimony sulfide, Sb_2S_3	16		
Sulfur	10		
Charcoal	10		
II. Potassium nitrate, KNO_3	55	Gold glitter	Used in aerial stars
"Bright" aluminum powder	5		
Dextrine	4		
Antimony sulfide, Sb_2S_3	14		
Charcoal	8		
Sulfur	8		
III. Potassium nitrate, KNO_3	55	Good white glitter	Used in fountains
Sulfur	10		
Charcoal	10		
Atomized aluminum	10		
Iron oxide, Fe_2O_3	5		
Barium carbonate, $BaCO_3$	5		
Barium nitrate, $Ba(NO_3)_2$	5		

[a]Reference 10.

blue and red light in the proper proportions will produce a purple effect. Color theory is a complex topic, but it is one that should be studied by anyone desiring to produce colored flames [2].

The production of a vividly-colored flame is a much more challenging problem than creating white light. A delicate balance of factors is required to obtain a satisfactory effect:

1. An atomic or molecular species that will emit the desired wavelength, or blend of wavelengths, must be present in the pyrotechnic flame.
2. The emitting species must be sufficiently volatile to exist in the vapor state at the temperature of the pyrotechnic

TABLE 7.6 The Visible Spectrum[a]

Wavelength (nanometers)	Emission color	Observed color — if this wavelength is removed from white light
<380	None (ultraviolet region)	—
380-435	Violet	Yellowish-green
435-480	Blue	Yellow
480-490	Greenish-blue	Orange
490-500	Bluish-green	Red
500-560	Green	Purple
560-580	Yellowish-green	Violet
580-595	Yellow	Blue
595-650	Orange	Greenish-blue
650-780	Red	Bluish-green
>780	None (infrared region)	—

[a]Source: H. H. Bauer, G. D. Christian, and J. E. O'Reilly, *Instrumental Analysis*, Allyn & Bacon, Inc., Boston, 1979.

reaction. The flame temperature will range from 1000-2000°C (or more), depending on the particular composition used.

3. Sufficient heat must be generated by the oxidizer/fuel reaction to produce the excited electronic state of the emitter. A minimum heat requirement of 0.8 kcal/gram has been mentioned by Shidlovskiy [5].
4. Heat is necessary to volatilize and excite the emitter, but you must not *exceed* the dissociation temperature of molecular species (or the ionization temperature of atomic species) or color quality will suffer. For example, the green emitter BaCl is unstable above 2000°C and the best blue emitter, CuCl, should not be heated above 1200°C [5].

A temperature *range* is therefore required, high enough to achieve the excited electronic state of the vaporized species but low enough to minimize dissociation.

5. The presence of incandescent solid or liquid particles in the flame will adversely affect color quality. The resulting "black body" emission of white light will enhance overall emission intensity, but the color quality will be lessened. A "washed out" color will be perceived by viewers. The use of magnesium or aluminum metal in color compositions will yield high flame temperatures and high overall intensity, but broad emission from incandescent magnesium oxide or aluminum oxide products may lower color purity.
6. Every effort must be made to minimize the presence of unwanted atomic and molecular emitters in the flame. Sodium compounds can not be used in any color mixtures except yellow. The strong yellow atomic emission from sodium (589 nanometers) will overwhelm other colors. Potassium emits weak violet light (near 450 nanometers), but good red and green flames can be produced with potassium compounds present in the mixture. Ammonium perchlorate is advantageous for color compositions because it contains no metal ion to interfere with color quality. The *best* oxidizer to choose, therefore, should contain the metal ion whose emission, in atomic or molecular form, is to be used for color production, *if* such an oxidizer is commercially available, works well, and is safe to use. Using this logic, the chemist would select barium nitrate or barium chlorate for green flame mixtures. Strontium nitrate, although hygroscopic, is frequently selected for red compositions. The use of a salt other than one with an oxidizing anion (e.g., strontium carbonate for red) may be required by hygroscopicity and safety considerations. However, these inert ingredients will tend to lower the flame temperature and therefore lower the emission intensity. A low percentage of color ingredient must be used in such cases to produce a satisfactory color.
7. If a binder is required in a colored flame mixture, the minimum possible percentage should be used. Carbon-containing compounds may be oxidized to the atomic carbon level in the flame and produce an orange color. The use of a binder that is already substantially oxidized (one with a high oxygen content, such as dextrine) can minimize this problem. Binders such as paraffin that contain little or no oxygen should be avoided unless a hot, oxygen-rich composition is being prepared.

Oxidizer Selection

The numerous requirements for a good oxidizer were discussed in detail in Chapter 3. An oxidizer for a colored flame composition must meet all of those requirements, and in addition must either emit the proper wavelength light to yield the desired color or *not* emit any light that interferes with the color produced by other components.

In addition, the oxidizer must react with the selected fuel to produce a flame temperature that yields the maximum emission of light in the proper wavelength range. If the temperature is too low, not enough "excited" molecules are produced and weak color intensity is observed. A flame temperature that is too hot may decompose the molecular emitter, destroying color quality.

Table 7.7 gives some data on flame temperatures obtained by Shimizu for oxidizer/shellac mixtures. Sodium oxalate was added to yield a yellow flame color and permit temperature measurement by the "line reversal" method [11].

The data in Table 7.7 show that potassium nitrate, with its highly endothermic heat of decomposition, produces significantly lower flame temperatures with shellac than the other three oxidizers. The yellow light intensity will be substantially less for the nitrate compositions.

To use potassium nitrate in colored flame mixtures, it is necessary to include magnesium as a fuel to raise the flame temperature. A source of chlorine is also needed for formation of volatile BaCl (green), or SrCl (red) emitters. The presence of chlorine in the flame also aids by hindering the formation of magnesium oxide and strontium or barium oxide, all of which will hurt the color quality. Shidlovskiy suggests a minimum of 15% chlorine donor in a color composition when magnesium metal is used as a fuel [5].

Fuels and Burning Rates

Applications involving colored flame compositions will require either a long-burning composition or a mixture that burns rapidly to give a burst of color.

Highway flares ("fusees") and the "lances" used to create fireworks set pieces require long burning times ranging from 1-30 minutes. "Fast" fuels such as metal powders and charcoal are usually *not* included in these slow mixtures. Partially-oxidized organic fuels such as dextrine can be used. Coarse oxidizer and fuel particles can also retard the burning rate. Highway flares

TABLE 7.7 Flame Temperatures for Oxidizer/Shellac Mixtures

Composition	Flame temperatures for various oxidizers (°C)[a]			
	Potassium perchlorate $KClO_4$	Ammonium perchlorate NH_4ClO_4	Potassium chlorate $KClO_3$	Potassium nitrate KNO_3
I. 75% Oxidizer 15% Shellac 10% Sodium oxalate[b]	2250	2200	2180	1675
II. 70% Oxidizer 20% Shellac 10% Sodium oxalate	2125	2075	2000	1700
III. 65% Oxidizer 25% Shellac 10% Sodium oxalate	1850	1875	1825	1725

[a]Reference 11.
[b]The sodium oxalate ($Na_2C_2O_4$) produces a yellow flame. The intensity of the yellow light emission can be used to determine the flame temperature.

often contain sawdust as a coarse, slow-burning retardant to help achieve lengthy burning times.

To achieve rapid burning – such as in the brightly-colored "stars" used in aerial fireworks and Very pistol cartridges – compositions will contain charcoal or a metallic fuel (usually magnesium). Fine particle sizes will be used, and all ingredients will be well-mixed to achieve a very homogeneous – and fast burning – mixture.

Color Intensifiers

Chlorine is the key to the production of good red, green, and blue flames, and its presence is required in a pyrotechnic mixture to

TABLE 7.8 Chlorine Donors for Pyrotechnic Mixtures

Material	Formula	Melting point, °C	% Chlorine by weight
Polyvinyl chloride	$(-CH_2CHCl-)_n$	Softens ca. 80 decomposes ca. 160	56
"Parlon" (chlorinated polyisopropylene)		Softens 140	ca. 66
Hexachlorobenzene	C_6Cl_6	229	74.7
"Dechlorane" (hexachloropentadiene dimer)	$C_{10}Cl_{12}$	160	78.3
Hexachloroethane	C_2Cl_6	185	89.9

achieve a good output of these colors. Chlorine serves *two* important functions in a pyrotechnic flame. It forms volatile chlorine-containing molecular species with the color-forming metals, ensuring a sufficient concentration of emitters in the vapor phase. Also, these chlorine-containing species are good emitters of narrow bands of visible light, producing the observed flame color. Without *both* of these properties – volatility and light emission – good colors would be difficult to achieve.

The use of chlorate or perchlorate oxidizers ($KClO_3$, $KClO_4$, etc.) is one way to introduce chlorine atoms into the pyrotechnic flame. Another method is to incorporate a chlorine-rich organic compound into the mixture. Table 7.8 lists some of the chlorine donors commonly used in pyrotechnic mixtures. A dramatic increase in color quality can be achieved by the addition of a small percentage of one of these materials into a mixture. Shimizu recommends the addition of 2-3% organic chlorine donor into compositions that don't contain a metallic fuel, and the addition of 10-15% chlorine donor into the high temperature mixtures containing metallic fuels [11].

Shimizu attributes much of the value of these chlorine donors in magnesium-containing compositions to the production in the flame of hydrogen chloride, which reacts with magnesium oxide to form volatile MgCl molecules. The incandescent emission from

MgO particles is thereby reduced, and color quality improves significantly.

$MgO + HCl \rightarrow MgCl + OH$

Red Flame Compositions

The best flame emission in the red region of the visible spectrum is produced by molecular strontium monochloride, SrCl. This species – unstable at room temperature – is generated in the pyrotechnic flame by a reaction between strontium and chlorine atoms. Strontium dichloride, $SrCl_2$, would appear to be a logical precursor to SrCl, and it is readily available commercially, but it is much too hygroscopic to use in pyrotechnic mixtures.

The SrCl molecule emits a series of bands in the 620-640 nanometer region – the "deep red" portion of the visible spectrum. Other peaks are observed. Strontium monohydroxide, SrOH, is another substantial emitter in the red and orange-red regions [1, 11]. The emission spectrum of a red flare is shown in Figure 7.1.

Strontium nitrate – $Sr(NO_3)_2$ – is often used as a combination oxidizer/color source in red flame mixtures. A "hotter" oxidizer, such as potassium perchlorate, is frequently used to help achieve higher temperatures and faster burning rates. Strontium nitrate is rather hygroscopic, and water can not be used to moisten a binder for mixtures using this oxidizer. Strontium carbonate is much less hygroscopic and can give a beautiful red flame under the proper conditions. However, it contains an inert anion – the carbonate ion, CO_3^{-2} — and low percentages must be used to avoid burning difficulties.

To keep the SrCl from oxidizing in the flame, Shidlovskiy recommends using a composition containing a negative oxygen balance (excess fuel). Such a mixture will minimize the reaction

$2\ SrCl + O_2 \rightarrow 2\ SrO + Cl_2$

and enhance color quality [5]. Several red formulas are presented in Table 7.9

Green Flame Compositions

Pyrotechnic compositions containing a barium compound and a good chlorine source can generate barium monochloride, BaCl, in the flame and the emission of green light will be observed. BaCl – an unstable species at room temperature – is an excellent emitter in

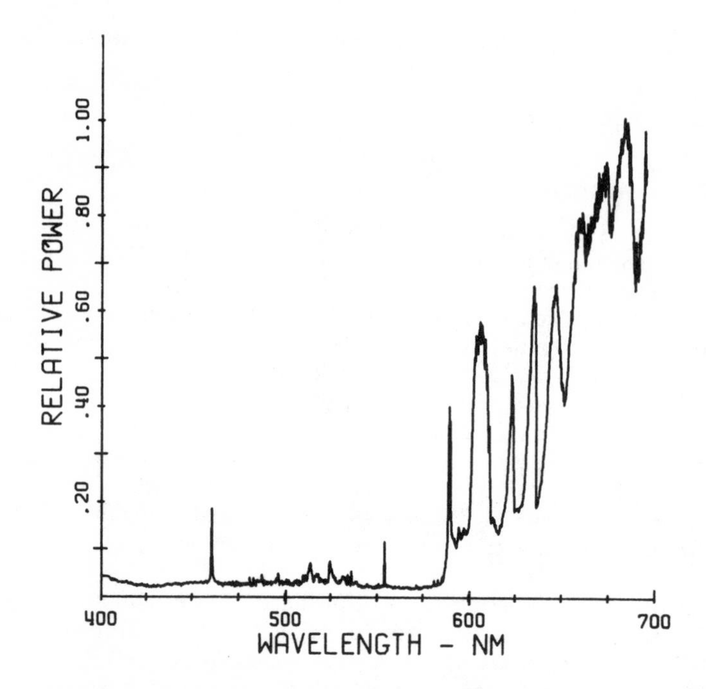

FIG. 7.1 Emission spectrum of a red flare. Emission is concentrated in the 600-700 nm region. The primary emitting species are SrCl and SrOH molecules in the vapor state. The composition of the flare was potassium perchlorate (20.5%), strontium nitrate (34.7%), magnesium (24.4%), polyvinylchloride (11.4%), and asphaltum (9.0%). Source: H. A. Webster III, "Visible Spectra of Standard Navy Colored Flares," Proceedings, Explosives and Pyrotechnics Applications Section, American Defense Preparedness Association, Fort Worth, Texas, September, 1983.

the 505-535 nanometer region of the visible spectrum – the "deep green" portion [1, 11]. The emission spectrum of a green flare was shown in Figure 4.1.

Barium nitrate — $Ba(NO_3)_2$ — and barium chlorate — $Ba(ClO_3)_2$ — are used most often to produce green flames, serving both as the oxidizer and color source. Barium chlorate can produce a deep green, but it is somewhat unstable and can form explosive mixtures with good fuels. Barium nitrate produces an acceptable green color, and it is considerably safer to work with due to its

TABLE 7.9 Red Flame Compositions

Composition	% by weight	Use	Reference
I. Ammonium perchlorate, NH_4ClO_4	70	Red torch	6
Strontium carbonate, $SrCO_3$	10		
Wood meal (slow fuel)	20		
II. Potassium perchlorate, $KClO_4$	67	Red fireworks star	6
Strontium carbonate, $SrCO_3$	13.5		
Pine root pitch	13.5		
Rice starch	6		
III. Potassium perchlorate, $KClO_4$	32.7	Red fireworks star	9
Ammonium perchlorate, NH_4ClO_4	28.0		
Strontium carbonate, $SrCO_3$	16.9		
Red gum	14.0		
Hexamethylenetetramine, $C_6H_{12}N_4$	2.8		
Charcoal	1.9		
Dextrine (dampen with 3:1 water/alcohol)	3.7		
IV. Potassium perchlorate, $KClO_4$	44	Red signal flare (very little residue)	Unpublished
Strontium nitrate, $Sr(NO_3)_2$	31		
Epoxy fuel/binder	25		

high decomposition temperature and endothermic heat of decomposition. Barium carbonate ($BaCO_3$) is another possibility, but it must be used in low percentage due to its inert anion, CO_3.

An oxygen-deficient flame is required for a good-quality green flame. Otherwise, barium oxide (BaO) will form and emit a series of bands in the 480-600 nanometer range, yielding a dull, yellowish-green color. The reaction

$$2\ BaCl + O_2 \rightarrow 2\ BaO + Cl_2$$

will shift to the left-hand side when chlorine is present in abundance and oxygen is scarce, and a good green color will be achieved. A flame temperature that is too high will decompose BaCl, however, so metal fuels must be held to a minimum, if they are used at all. A "cool" flame is best.

This temperature dependence and need for chlorine source are important to remember. A binary mixture of barium nitrate and magnesium metal will produce a brilliant *white* light upon ignition, from a combination of MgO and BaO emission at the high temperature achieved by the mixture. Addition of a chlorine-containing organic fuel to lower the temperature and provide chlorine atoms to form BaCl can produce a green flame. Several green flame compositions are given in Table 7.10.

Blue Flame Compositions

The generation of an intense, deep-blue flame represents the ultimate challenge to the pyrotechnic chemist. A delicate balance of temperature and molecular behavior is required to obtain a good blue, but it can be done if the conditions are right.

The best flame emission in the blue region of the visible spectrum (435-480 nanometers) is obtained from copper monochloride, CuCl. Flame emission from this molecular species yields a series of bands in the region from 428-452 nanometers, with additional peaks between 476-488 nanometers [1, 11].

In an oxygen-rich flame, and at temperatures above 1200°C, CuCl is unstable and will react to form CuO and CuOH. CuOH emits in the 525-555 nanometer region (green!) and substantial emission may overpower any blue effect that is also present. Copper oxide, CuO, emits a series of bands in the red region, and this reddish emission is often seen at the top of blue flames, where sufficient oxygen from the atmosphere is present to convert CuCl to CuO [11].

Paris green – copper acetoarsenite, $(CuO)_3As_2O_3Cu(C_2H_3O_2)$ — was widely used in blue flame mixtures until a few years ago. It produces a good blue flame, but it has all but vanished from commercial formulas because of the health hazards associated with its use. (It contains arsenic!)

Copper oxide (CuO), basic copper carbonate – $CuCO_3 \cdot Cu(OH)_2$, and copper sulfate – available commercially as $CuSO_4 \cdot 5H_2O$ – are

TABLE 7.10 Green Flame Compositions

Composition	% by weight	Use	Reference
I. Ammonium perchlorate, NH_4ClO_4	50	Green torch	6
Barium nitrate, $Ba(NO_3)_2$	34		
Wood meal	8		
Shellac	8		
II. Barium chlorate, $Ba(ClO_3)_2 \cdot H_2O$	65	Green torch	Unpublished
Barium nitrate, $Ba(NO_3)_2$	25		
Red gum	10		
III. Potassium perchlorate, $KClO_4$	46	Green fireworks star	6
Barium nitrate, $Ba(NO_3)_2$	32		
Pine root pitch	16		
Rice starch	6		
IV. Barium nitrate, $Ba(NO_3)_2$	59	Russian green fire	5
Polyvinyl chloride	22		
Magnesium	19		

among the materials used in blue flame mixtures. Potassium perchlorate and ammonium perchlorate are the oxidizers found in most blue compositions. Potassium chlorate would be an ideal choice because of its ability to sustain reaction at low temperatures (remember, CuCl is unstable above 1200°C), but copper chlorate is an extremely reactive material. The chance of it forming should a blue mixture get wet precludes the commercial use of $KClO_3$.

Several formulas for blue flame compositions are given in Table 7.11. An extensive review of blue and purple flames, concentrating on potassium perchlorate mixtures, has been published by Shimizu [13].

TABLE 7.11 Blue Flame Compositions

Compositions	% by weight	Use	Reference
I. Potassium perchlorate, $KClO_4$	68.5	Blue flame-"excellent"	13
Polyvinyl chloride	9		
Copper oxide, CuO	15		
Red gum	7.5		
Rice starch	5 (additional %)		
II. Potassium perchlorate, $KClO_4$	40	Blue flame	6
Ammonium perchlorate, NH_4ClO_4	30		
Copper carbonate, $CuCO_3$[a]	15		
Red gum	15		
III. Potassium perchlorate, $KClO_4$	68	Blue flame-"excellent"	13
Copper carbonate, $CuCO_3$[a]	15		
Polyvinyl chloride	11		
Red gum	6		
Rice starch	5 (additional %)		
IV. Ammonium perchlorate, NH_4ClO_4	70	Blue fireworks star (with charcoal tail)	14
Red gum	10		
Copper carbonate, $CuCO_3$[a]	10		
Charcoal	10		
Dextrine (moisten with iso-propyl alcohol)	5 (additional %)		

[a]Material is actually basic copper carbonate, $2\ CuCO_3 \cdot Cu(OH)_2$

TABLE 7.12 Purple Flame Compositions

Composition	% by weight	Comment[a]
I. Potassium perchlorate, $KClO_4$	70	"Excellent"
Polyvinyl chloride	10	
Red gum	5	
Copper oxide, CuO	6	
Strontium carbonate, $SrCO_3$	9	
Rice starch	5 (additional %)	
II. Potassium perchlorate, $KClO_4$	70	"Excellent"
Polyvinyl chloride	10	
Red gum	5	
Copper powder, Cu	6	
Strontium carbonate, $SrCO_3$	9	
Rice starch	5 (additional %)	

[a]Reference 13.

Purple Flame Compositions

A purple flame, a relative newcomer to pyrotechnics, can be achieved by the correct balance of red and blue emitters. The additive blending of these two colors produces a perception of purple by an observer. Several comprehensive review articles on purple flames have recently been published [13].

The compositions given in Table 7.12 received an "excellent" rating in the review article written by Shimizu [13].

Yellow Flame Compositions

Yellow flame color is achieved by atomic emission from sodium. The emission intensity at 589 nanometers increases as the reaction temperature is raised; there is no molecular emitting species here to decompose. Ionization of sodium atoms to sodium ions will occur at very high temperatures, however, so even here there is an upper limit of temperature that must be avoided for maximum color quality. The emission spectrum of a yellow flare is shown in Figure 7.2.

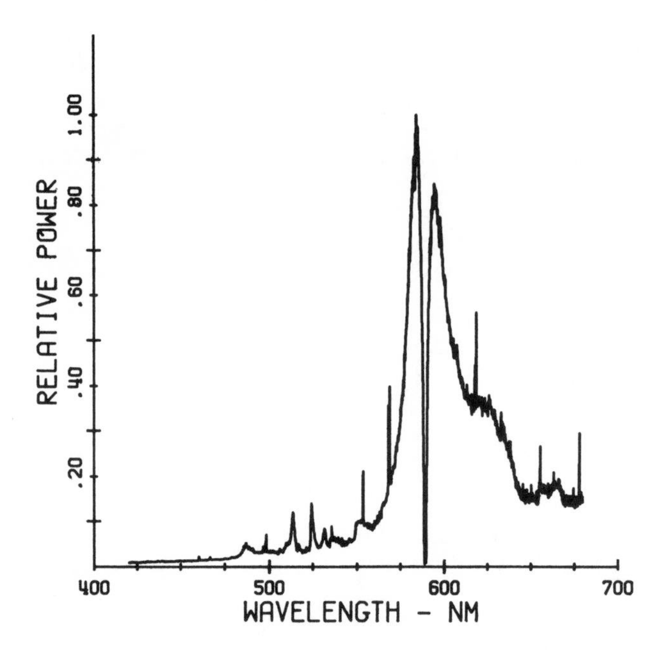

FIG. 7.2 Emission spectrum of a yellow flare. The primary emitting species is atomic sodium, with intensity centered near 589 nm. A background continuum of "blackbody" emission and bands from vaporized BaO, BaOH, and BaCl are also observed. The composition of the flare was potassium perchlorate (21.0%), barium nitrate (20.0%), magnesium (30.3%), sodium oxalate (19.8%), asphaltum (3.9%), and binder (5.0%). This is apparently a former green flare formula to which sodium oxalate was added to obtain a yellow flame. The intense atomic sodium emission at 589 nm overwhelms the green bands from barium-containing species! Source: H. A. Webster III, "Visible Spectra of Standard Navy Colored Flares," Proceedings, Explosives and Pyrotechnics Applications Section, American Defense Preparedness Association, Fort Worth, Texas, September, 1983.

Most sodium compounds tend to be quite hygroscopic, and therefore simple compounds such as sodium nitrate ($NaNO_3$), sodium chlorate ($NaClO_3$), and sodium perchlorate ($NaClO_4$) – combining the oxidizing anion with the metallic emitter – can not be used unless precautions are taken to protect against moisture

TABLE 7.13 Yellow Flame Compositions

Composition	% by weight	Use	Reference
I. Potassium perchlorate, $KClO_4$	70	Yellow fireworks star	6
Sodium oxalate, $Na_2C_2O_4$	14		
Red gum	6		
Shellac	6		
Dextrine	4		
II. Potassium perchlorate, $KClO_4$	75	Yellow fire	6
Cryolite, Na_3AlF_6	10		
Red gum	15		
III. Sodium nitrate, $NaNO_3$	56	Yellow fire (Russian)	5
Magnesium	17		
Polyvinyl chloride	27		
IV. Potassium nitrate, KNO_3	37	Yellow fire (Russian)	5
Sodium oxalate, $Na_2C_2O_4$	30		
Magnesium	30		
Resin	3		
V. Barium nitrate, $Ba(NO_3)_2$	17	Yellow flare	8
Strontium nitrate, $Sr(NO_3)_2$	16		
Potassium perchlorate, $KClO_4$	17		
Sodium oxalate, $Na_2C_2O_4$	17		
Hexachlorobenzene, C_6Cl_6	12		
Magnesium	18		
Linseed oil	3		

before, during, and after the manufacturing process. Sodium oxalate ($Na_2C_2O_4$) and cryolite (Na_3AlF_6) are low in hygroscopicity and they are therefore the color agents used in most commercial yellow flame mixtures. Some representative yellow compositions are given in Table 7.13.

REFERENCES

1. B. E. Douda, "Theory of Colored Flame Production," RDTN No. 71, U.S. Naval Ammunition Depot, Crane, Indiana, 1964.

2. K. L. Kosanke, "The Physics, Chemistry and Perception of Colored Flames," *Pyrotechnica VII*, Pyrotechnica Publications, Austin, Texas, 1981.
3. B. E. Douda, "Spectral Observations in Illuminating Flames," *Proceedings, First International Pyrotechnics Seminar,* Denver Research Institute, Estes Park, Colorado, August, 1968, p. 113 (available from NTIS as AD 679 911).
4. D. R. Dillehay, "Pyrotechnic Flame Modeling for Sodium D-Line Emissions," *Proceedings, Fifth International Pyrotechnics Seminar*, Denver Research Institute, Vail, Colorado, July, 1976, p. 123 (available from NTIS as AD A087 513).
5. A. A. Shidlovskiy, *Principles of Pyrotechnics*, 3rd Ed., Moscow, 1964. (Translated by Foreign Technology Division, Wright-Patterson Air Force Base, Ohio, 1974.)
6. T. Shimizu in R. Lancaster's *Fireworks Principles and Practice*, Chemical Publishing Co., Inc., New York, 1972.
7. U.S. Army Material Command, Engineering Design Handbook, Military Pyrotechnic Series, Part One, "Theory and Application," Washington, D.C., 1967 (AMC Pamphlet 706-185).
8. F. L. McIntyre, "A Compilation of Hazard and Test Data for Pyrotechnic Compositions," Report ARLCD-CR-80047, U.S. Army Armament Research and Development Command, Dover, NJ, 1980.
9. *Pyrotechnica IV*, Pyrotechnica Publications, Austin, Texas, 1978.
10. R. M. Winokur, "The Pyrotechnic Phenomenon of Glitter," *Pyrotechnica II*, Pyrotechnica Publications, Austin, Texas, 1978.
11. T. Shimizu, *Fireworks – The Art, Science and Technique*, pub. by T. Shimizu, distrib. by Maruzen Co., Ltd., Tokyo, 1981.
12. T. Shimizu, "Studies on Strobe Light Pyrotechnic Compositions," *Pyrotechnica VIII*, Pyrotechnica Publications, Austin, Texas, 1982.
13. T. Shimizu, "Studies on Blue and Purple Flame Compositions Made With Potassium Perchlorate," *Pyrotechnica VI*, Pyrotechnica Publications, Austin, Texas, 1980.
14. *Pyrotechnica I*, Pyrotechnica Publications, Austin, Texas, 1977.

A portion of the "finale" of a fireworks display. Several hundred aerial shells are usually launched in a brief period of time to overwhelm the senses of the audience. A Japanese "chrysanthemum" shell with its characteristic large, symmetrical burst of color can be seen near the center of the photograph. Several American aerial shells, with their more-random bursting pattern, can also be seen. The bright "dots" of light seen in the picture are the bursts of "salutes"; these are tubes containing "flash and sound" composition that explode to create a booming noise and a flash of light. (Zambelli Internationale)

8
SMOKE AND SOUND

SMOKE PRODUCTION

Most explosive and pyrotechnic reactions produce significant quantities of smoke, and this visible phenomenon may or may not be desirable. Smoke can obscure colored flames, and therefore attempts are made to keep the production of smoke to a minimum in such mixtures. However, a variety of smoke-producing compositions are purposefully manufactured for use in daytime signalling and troop and equipment obscuration, as well as for amusement and entertainment purposes.

Two basic processes are used to create smoke clouds: the condensation of vaporized material and the dispersion of solid or liquid particles. Materials can either be released slowly via a pyrotechnic reaction or they can instantaneously be scattered using an explosive material. Technically, a dispersion of fine solid particles in air is termed a *smoke*, while liquid particles in air create a *fog*. A smoke is created by particles in the 10^{-5}-10^{-9} meter range, while larger suspended particles create a *dust* [1].

A variety of events that will lead to smoke production can occur in the pyrotechnic flame. Incomplete burning of an organic fuel will produce a black, sooty flame (mainly atomic carbon). A highly-oxidized fuel such as a sugar is not likely to produce carbon. Materials such as naphthalene ($C_{10}H_8$) and anthracene ($C_{14}H_{10}$) – volatile solids with high carbon content – are good candidates for soot production. Several mixtures that will produce black smokes are listed in Table 8.1.

The heat from the reaction between an oxidizer and fuel can vaporize a volatile ingredient, with no chemical change occurring

TABLE 8.1 Black Smoke Compositions

Composition	% by weight	Reference
I. Potassium chlorate, $KClO_3$	55	1
Anthracene, $C_{14}H_{10}$	45	
II. Potassium chlorate, $KClO_3$	45	1
Naphthalene, $C_{10}H_8$	40	
Charcoal	15	
III. Potassium perchlorate, $KClO_4$	56	2
Sulfur	11	
Anthracene, $C_{14}H_{10}$	33	
IV. Hexachloroethane, C_2Cl_6	62	2
Magnesium	15	
Naphthalene (or anthracene)	23	

in the vaporized material. The vaporized component, which was part of the original mixture, then condenses as fine, solid particles upon leaving the reaction zone and a smoke is created. Organic dyes, ammonium chloride, and sulfur can be used to create smokes using this method.

Alternately, the pyrotechnic reaction can occur in a separate container, and the heat that is produced volatilizes a smoke-forming component contained in an adjacent compartment. The vaporization and dispersion of heavy oils to create white smoke uses this technique.

Finally, a *product* of a pyrotechnic reaction may vaporize from the reaction zone and subsequently condense as fine particles in air, creating a smoke. Potassium chloride (boiling point 1407°C) produces smoke in many potassium chlorate and potassium perchlorate compositions, although smoke is usually not a goal sought from these mixtures.

A good white smoke can be obtained by the formation of zinc chloride, $ZnCl_2$, from a reaction between zinc metal and a chlorinated organic compound (the chlorine-containing species serves as the oxidizer). Reaction products that strongly attract moisture (such as $ZnCl_2$) will have an enhanced smoke effect in humid atmospheres. The burning of elemental phosphorus, producing phosphorus oxides, creates dense white smoke as the oxides attract moisture to form acids such as phosphoric acid, H_3PO_4.

COLORED SMOKE MIXTURES

The generation of colored smoke by the volatilization of an organic dye is a fascinating pyrotechnic problem. The military and the fireworks and entertainment industries rely on this technique for the generation of copious quantities of brilliantly-colored smoke.

The requirements for an effective colored-smoke composition include:

1. The mixture must produce sufficient heat to vaporize the dye, as well as produce a sufficient volume of gas to disperse the dye into the surrounding space.
2. The mixture must ignite at a low temperature and continue to burn smoothly at low temperature (well below 1000°C). If the temperature is too high, the dye molecules will decompose and the color quality as well as volume of the smoke will deteriorate. Metal fuels are not used in colored smoke mixtures because of the high reaction temperatures they produce.
3. Although a low ignition temperature is required, the smoke mixture must be *stable* during manufacturing and storage, over the expected range of ambient temperatures.
4. The molecules creating the colored smoke must be of low toxicity (including low carcinogenicity). Further, they must readily sublime without decomposition at the temperature of the pyrotechnic reaction to yield a dense smoke of good color quality [3].

When requirements that include low ignition temperature and reliable propagation of burning at low reaction temperature are considered, the choice of oxidizer rapidly narrows to one candidate – potassium chlorate, $KClO_3$. The ignition temperature of potassium chlorate combined with sulfur or many organic fuels is below 250°C. Good heat production is achieved with such mixtures, in part due to the exothermic decomposition of $KClO_3$ at a temperature below 400°C, forming KCl and oxygen gas.

A mixture consisting of 70% $KClO_3$ and 30% sugar ignites at 220°C and has a heat of reaction of approximately 0.8 kcal/gram

[5]. Both chlorate-sulfur and chlorate-sugar mixtures are used in commercial colored smoke compositions. Sodium bicarbonate ($NaHCO_3$) is added to $KClO_3$/S mixtures to neutralize any acidic impurities that might stimulate premature ignition of the composition, and it also acts as a coolant by decomposing endothermically to evolve carbon dioxide gas (CO_2). Magnesium carbonate ($MgCO_3$) is also used as a coolant, absorbing heat to decompose into magnesium oxide (MgO) and CO_2. The amount of coolant can be used to help obtain the desired rate of burning and the correct reaction temperature – if a mixture burns too rapidly, more coolant should be added.

The ratio of oxidizer to fuel will also affect the amount of heat and gas that are produced. A stoichiometric mixture of $KClO_3$ and sulfur (equation 8.1) contains a 2.55:1 ratio of oxidizer to fuel, by weight. Colored smoke mixtures in use today contain ratios very close to this stoichiometric amount. The chlorate/sulfur reaction is not strongly exothermic, and a stoichiometric mixture is needed to generate the heat necessary to volatilize the dye.

$$2\,KClO_3 + 3\,S \rightarrow 3\,SO_2 + 2\,KCl \qquad (8.1)$$

grams 245 96

% 71.9 28.1 (a 2.55 to 1.00 ratio)

The reaction of potassium chlorate with a carbohydrate (e.g., lactose) will produce carbon monoxide (CO), carbon dioxide (CO_2), or a mixture depending on the oxidizer:fuel ratio. The balanced equations are given as equations 8.2 and 8.3. (Lactose occurs as a hydrate – one water molecule crystallizes with each lactose molecule.)

<u>CO_2 Product:</u>

$$8\,KClO_3 + C_{12}H_{22}O_{11}\cdot H_2O \rightarrow 8\,KCl + 12\,CO_2 + 12\,H_2O \qquad (8.2)$$

grams 980 360.3

% 73.1 26.9 (2.72 to 1.00 ratio)

Heat of reaction = 1.06 kcal/gram [1]

<u>CO Product:</u>

$$4\,KClO_3 + C_{12}H_{22}O_{11}\cdot H_2O \rightarrow 4\,KCl + 12\,CO + 12\,H_2O \qquad (8.3)$$

grams 490 360.3

% 57.6 42.4 (1.36 to 1.00 ratio)

Heat of reaction = 0.63 kcal/gram [1]

TABLE 8.2 Colored Smoke Compositions

Composition	% by weight	Reference
Green smoke		
Potassium chlorate, $KClO_3$	25.4	8
Sulfur	10.0	
Green dye	40.0	
Sodium bicarbonate, $NaHCO_3$	24.6	
Red smoke		
Potassium chlorate, $KClO_3$	29.5	8
Lactose	18.0	
Red dye	47.5	
Magnesium carbonate, $MgCO_3$	5.0	
New yellow smoke		
Potassium chlorate, $KClO_3$	22.0	4
Sucrose	15.0	
Chinoline yellow dye	42.0	
Magnesium carbonate, $MgCO_3$	21.0	

The amount of heat can be controlled by adjusting the $KClO_3$: sugar ratio. Excess oxidizer should be avoided; it will encourage oxidation of the dye molecules. The quantity (and volatility) of the dye will also affect the burning rate. The greater the quantity of dye used, the slower will be the burning rate – the dye is a diluent in these mixtures. Typical colored smoke compositions contain 40-60% dye by weight. Table 8.2 shows a variety of colored smoke compositions.

In colored smoke compositions, the volatile organic dye sublimes out of the reacting mixture and then condenses in air to form small solid particles. The dyes are strong *absorbers* of visible light. The light that is reflected off these particles is *missing* the absorbed wavelengths, and the *complementary* hue is perceived by observers. This color-producing process is different from that of colored flame production, where the *emitted* wavelengths are perceived as color by viewers. Table 7.6 lists the complementary colors for the various regions of the visible spectrum.

A variety of dyes have been used in colored smoke mixtures; many of these dyes are presently under investigation for carcinogenicity and other potential health hazards because of their molecular similarity to known "problem" compounds [4]. The materials that work best in colored smokes have several properties in common, including:

1. *Volatility*: The dye must convert to the vapor state on heating, without substantial decomposition. Only low molecular weight species (less than 400 grams/mole) are usually used – volatility typically decreases as molecular weight increases. Salts do not work well; ionic species generally have low volatility due to the strong inter-ionic attractions present in the crystalline lattice. Therefore, functional groups such as $-COO^-$ (carboxylate ion) and $-NR_3^+$ (a substituted ammonium salt) can not be present.
2. *Chemical stability*: Oxygen-rich functional groups ($-NO_2$, $-SO_3H$) can't be present. At the typical reaction temperatures of smoke compositions, these groups are likely to release their oxygen, leading to oxidative decomposition of the dye molecules. Groups such as $-NH_2$ and $-NHR$ (amines) are used, but one must be cautious of possible oxidative coupling reactions that can occur in an oxygen-rich environment.

Structures for some of the dyes used in colored smoke mixtures are given in Table 8.3.

WHITE SMOKE PRODUCTION

The processes used to generate a white smoke by means of a pyrotechnic reaction include:

1. *Sublimation of sulfur, using potassium nitrate as the oxidizer*: A 1:1 ratio of sulfur to KNO_3 is used in such mixtures. Caution: some toxic sulfur dioxide gas will be formed. Ignition of these mixtures must be done in a well-ventilated area.
2. *Combustion of phosphorus*: White or red phosphorus burns to produce various oxides of phosphorus, which then attract moisture to form dense white smoke. Research and

TABLE 8.3 Dyes for Colored Smoke Mixtures

Orange 7 α-xylene-azo-β-naphthol	Solvent green 3 1,4-di-p-toluidino-anthraquinone
Disperse red 9 1-methylamino-anthraquinone	Violet 1,4-diamino-2,3-dihydroanthraquinone
Chinoline yellow 2-(2-quinolyl)-1,3-indandione	Vat yellow 4 dibenzo(a,h)pyrene-7,14-dione

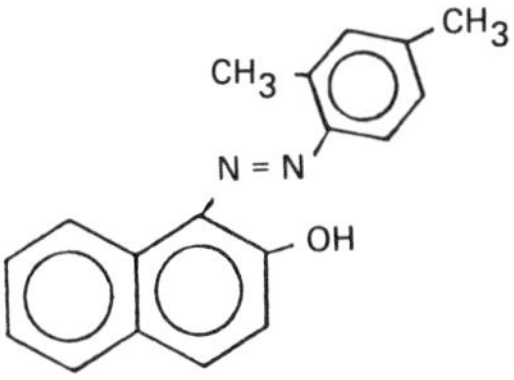

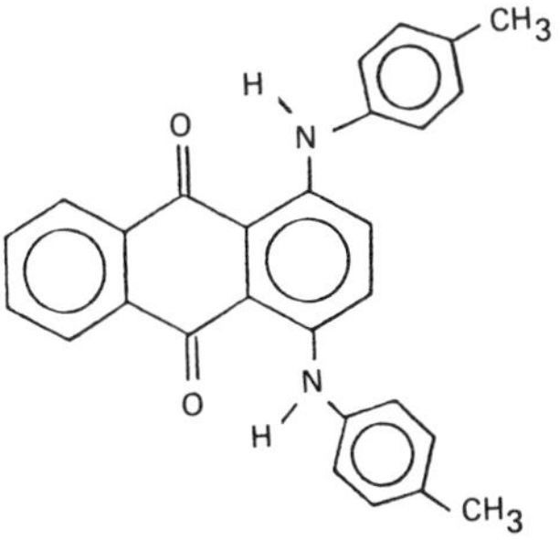

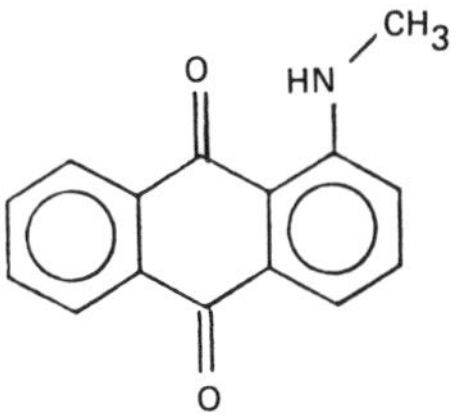

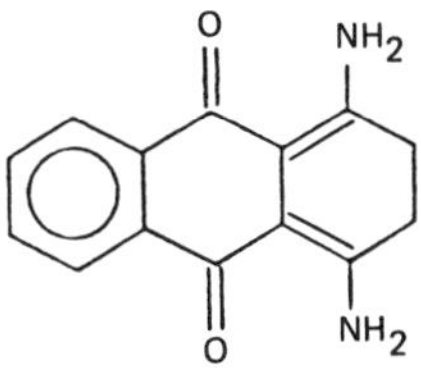

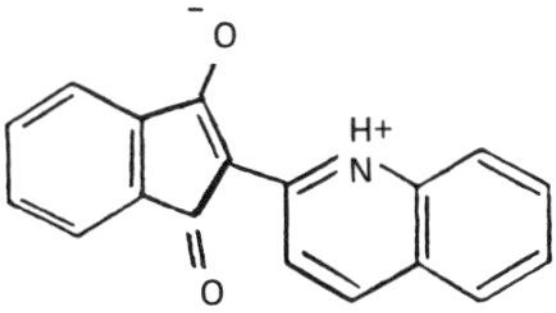

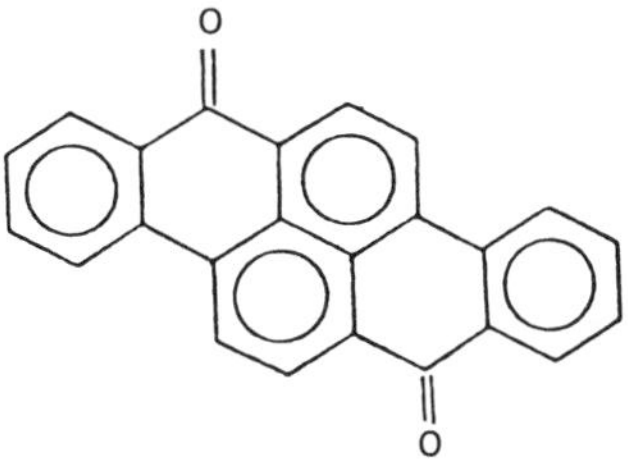

TABLE 8.4 White Smoke Compositions

Composition	% by weight	Note	Reference
I. Hexachloroethane, C_2Cl_6	45.5	HC type C	6
Zinc oxide, ZnO	47.5		
Aluminum	7.0		
II. Hexachlorobenzene, C_6Cl_6	34.4	Modified HC	6
Zinc oxide, ZnO	27.6		
Ammonium perchlorate, NH_4ClO_4	24.0		
Zinc dust	6.2		
Laminac	7.8		
III. Red phosphorus	63	Under development	8
Butyl rubber, methylene chloride	37		
IV. Red phosphorus	51.0		4
Magnesium	10.5		
Manganese dioxide, MnO_2	32.0		
Magnesium oxide, MgO	1.5		
Microcrystalline wax	5.0		
V. Potassium nitrate, KNO_3	48.5	Contains arsenic	9
Sulfur	48.5		
Arsenic disulfide, As_2S_2	3.0		

development work relating to red phosphorus-based smoke mixtures is actively being pursued to find substitutes for the zinc chloride smokes. A typical red phosphorus mixture is given in Table 8.4. An explosive bursting charge is often used with the very-hazardous white phosphorus. Caution: Phosphorus-based smokes generate acidic compounds which may be irritating to the eyes, skin, and respiratory tract.

3. *Volatilization of oil*: A pyrotechnic reaction produces the heat needed to vaporize high molecular weight hydrocarbons. The subsequent condensation of this oil in air creates a white smoke cloud. The toxicity of this smoke is probably the least of all the materials discussed here.

4. *Formation of zinc chloride ("HC Smokes")*: A reaction of the type

$$C_xCl_y + y/2\ Zn \rightarrow x\ C + y/2\ ZnCl_2 + heat$$

produces the zinc chloride vapor, which condenses in air and attracts moisture to create an effective white smoke. These mixtures have been widely used for over forty years with an excellent safety record during the manufacturing process. However, $ZnCl_2$ can cause headaches upon continued exposure and replacements for the HC smokes are actively being sought due to health concerns relating to the various reaction products.

The original HC smoke mixtures (Type A) contained zinc metal and hexachloroethane, but this composition is extremely moisture-sensitive and can ignite spontaneously if moistened. An alternative approach involves adding a small amount of aluminum metal to the composition, and zinc oxide (ZnO) is used in place of the moisture-sensitive metal. Upon ignition, a sequence of reactions ensues of the type [6]:

$$2\ Al + C_2Cl_6 \rightarrow 2\ AlCl_3 + 2\ C \qquad (8.4)$$

$$2\ AlCl_3 + 3\ ZnO \rightarrow 3\ ZnCl_2 + Al_2O_3 \qquad (8.5)$$

$$ZnO + C \rightarrow Zn + Co \qquad (8.6)$$

$$3\ Zn + C_2Cl_6 \rightarrow 3\ ZnCl_2 + 2\ C \qquad (8.7)$$

Alternatively, the original reaction has been proposed to be [7]:

$$2\ Al + 3\ ZnO \rightarrow 3\ Zn + Al_2O_3 \qquad (8.8)$$

In either event, the products are $ZnCl_2$, CO, and Al_2O_3. The zinc oxide cools and whitens the smoke by consuming atomic carbon in an endothermic reaction that occurs spontaneously above 1000°C (equation 8.6). The reaction with aluminum (equation 8.4 or 8.8) is quite exothermic, and this heat evolution controls the burning rate of the smoke mixture. A minimum amount of aluminum metal will yield the best white smoke. Several "HC" smoke compositions are listed in Table 8.4.

5. *"Cold Smoke"*: White smoke can also be achieved by nonthermal means. A beaker containing concentrated hydrochloric acid placed near a beaker of concentrated ammonia will generate white smoke by the vapor-phase reaction

$$HCl\ (gas) + NH_3\ (gas) \rightarrow NH_4Cl\ (solid)$$

Similarly, titanium tetrachloride ($TiCl_4$) rapidly reacts with moist air to produce a heavy cloud of titanium hydroxide – $Ti(OH)_4$ – and HCl.

NOISE

Two basic audible effects are produced by explosive and pyrotechnic devices: a loud explosive noise (called a "report" or "salute" in the fireworks industry) and a whistling sound.

A report is produced by igniting an explosive mixture, usually under confinement in a heavy-walled cardboard tube. Potassium chlorate and potassium perchlorate are the most commonly used oxidizers for report compositions, which are also referred to as "flash and sound" mixtures. These mixtures produce a flash of light and a loud "bang" upon ignition. Black powder under substantial confinement also produces a report.

"Flash and sound" compositions are true explosives, and they will detonate if a sufficient quantity of powder (perhaps 100 grams or more) is present in bulk form, even if unconfined! Chlorate-based mixtures are considerably more hazardous than perchlorate compositions because of their substantially lower ignition temperatures. However, flash and sound compositions made with *either* oxidizer must be considered *very* dangerous. They have killed many people at fireworks manufacturing plants in the United States and abroad. Mixing should only be done using remote means, and the smallest feasible amount of composition should be prepared at one time. Bulk flash and sound powder must *never* be stored anywhere near operating personnel.

The famous Chinese firecracker uses a mixture of potassium chlorate, sulfur, and aluminum. The chlorate combined with sulfur makes this mixture *doubly* dangerous for the manufacturer. The ignition temperature of the potassium chlorate/sulfur system is less than 200°C! The presence of aluminum – an excellent fuel – guarantees that the pyrotechnic reaction will rapidly propagate once it begins. Safety data from China is unavailable, but one has to wonder how many accidents occur annually from the preparation of this firecracker composition. The preparation of potassium chlorate/sulfur compositions was banned in Great Britain in 1894 because of the numerous accidents associated with this mixture!

TABLE 8.5 "Flash and Sound" Compositions[a]

Composition	% by weight	Use	Reference
I. Potassium perchlorate, $KClO_4$	50	Military simulator	8
Antimony sulfide, Sb_2S_3	33		
Magnesium	17		
II. Potassium perchlorate, $KClO_4$	64	M-80 firecracker for military training	8
Aluminum	22.5		
Sulfur	10		
Antimony sulfide, Sb_2S_3	3.5		
III. Potassium chlorate, $KClO_3$	43	Japanese "flash thunder" for aerial fireworks	5
Sulfur	26		
Aluminum	31		
IV. Potassium perchlorate, $KClO_4$	50	Japanese "flash thunder" for aerial fireworks	5
Sulfur	27		
Aluminum	23		

[a]*Note*: These mixtures are explosive and *very* dangerous. They must only be prepared by trained personnel using adequate protection, and should be mixed by remote means.

The standard American flash and sound composition is a blend of potassium perchlorate, sulfur or antimony sulfide, and aluminum. The ignition temperature of this formulation is several hundred degrees higher than chlorate-based mixtures, but these are *still* very dangerous compositions because of their extreme sensitivity to spark and flame. Ignition of a small portion of a "flash and sound" mixture will rapidly propagate through the entire sample. These mixtures should only be prepared remotely, by experienced personnel. Table 8.5 lists several "flash and sound" formulas.

TABLE 8.6 Whistle Compositions[a]

Composition	% by weight	Note	Reference
I. Potassium perchlorate, $KClO_4$	73	Military simulator	8
Gallic Acid, $C_7H_6O_5 \cdot H_2O$	24		
Red gum	3		
II. Potassium perchlorate, $KClO_4$	70	Perhaps the safest to prepare and use	5
Potassium benzoate, $KC_7H_5O_2$	30		
III. Potassium perchlorate, $KClO_4$	75	Hygroscopic-does not store well	5
Sodium salicylate, $NaC_7H_5O_3$	25		
IV. Potassium perchlorate, $KClO_4$	75	Chinese whistle composition	Unpublished
Potassium hydrogen phthalate, $KC_8H_5O_4$	25		

[a]*Note*: These mixtures are very sensitive to ignition and can be quite dangerous to prepare. They should only be mixed by trained personnel using adequate protection.

Whistles

A unique, whistling phenomenon can be produced by firmly pressing certain oxidizer/fuel mixtures into cardboard tubes and igniting the compositions. A detailed analysis of this phenomenon, both from a chemical and physical view, has been published by Maxwell [10].

A reaction that produces a whistling effect is burning intermittently from layer to layer in the pressed composition. A whistling reaction is on the verge of an explosion, so these mixtures must be cautiously prepared and carefully loaded into tubes. Large quantities of bulk powder should be avoided, and they should never be stored near operating personnel. Several formulas for whistle compositions are given in Table 8.6.

REFERENCES

1. A. A. Shidlovskiy, *Principles of Pyrotechnics*, 3rd Ed., Moscow, 1964. (Translated by Foreign Technology Division, Wright-Patterson Air Force Base, Ohio, 1974.)
2. T. Shimizu in R. Lancaster's *Fireworks Principles and Practice*, Chemical Publishing Co., Inc., New York, 1972.
3. A. Chin and L. Borer, "Investigations of the Effluents Produced During the Functioning of Navy Colored Smoke Devices," *Proceedings, Eighth International Pyrotechnics Seminar*, IIT Research Institute, Steamboat Springs, Colorado, July, 1982, p. 129.
4. M. D. Smith and F. M. Stewart, "Environmentally Acceptable Smoke Munitions," *Proceedings, Eighth International Pyrotechnics Seminar*, IIT Research Institute, Steamboat Springs, Colorado, July, 1982, p. 623.
5. T. Shimizu, *Fireworks – The Art, Science and Technique*, pub. by T. Shimizu, distrib. by Maruzen Co., Ltd., Tokyo, 1981.
6. U.S. Army Material Command, Engineering Design Handbook, Military Pyrotechnic Series, Part One, "Theory and Application," Washington, D.C., 1967 (AMC Pamphlet 706-185).
7. J. H. McLain, *Pyrotechnics from the Viewpoint of Solid State Chemistry*, The Franklin Institute Press, Philadelphia, Penna., 1980.
8. F. L. McIntyre, "A Compilation of Hazard and Test Data for Pyrotechnic Compositions," Report ARLCD-CR-80047, U.S. Army Armament Research and Development Command, Dover, NJ, 1980.
9. R. Lancaster, *Fireworks Principles and Practice*, Chemical Publishing Co., Inc., New York, 1972.
10. W. R. Maxwell, "Pyrotechnic Whistles," *4th Symposium on Combustion*, Williams and Wilkins, Baltimore, Md., 1953, p. 906.

APPENDIXES

APPENDIX A: OBTAINING PYROTECHNIC LITERATURE

Many of the technical reports and publications referenced in this book are available through the U.S. Department of Commerce's National Technical Information Service (NTIS) located in Springfield, Virginia.

Publications can be ordered from NTIS if the "accession numbers" are known; these are the numbers assigned by NTIS to technical documents in their files. NTIS can supply you with an "accession number" if you provide them with the title and author of a document. Current prices, order forms, accession numbers, and other needed information can be obtained from:

National Technical Information Service
5285 Port Royal Road
Springfield, VA 22161

NTIS numbers for several of the major references used in this book are:

A. A. Shidlovskiy, *Principles of Pyrotechnics*, 3rd Edition. NTIS # AD-A001859

Military Pyrotechnic Series, Part I, "Theory and Application." NTIS # AD-817071

Military Pyrotechnic Series, Part III, "Properties of Materials Used in Pyrotechnic Compositions." NTIS # AD-830394
F. L. McIntyre, "A Compilation of Hazard and Test Data for Pyrotechnic Compositions." NTIS # AD-A096248

In addition, copies of the various Proceedings of the International Pyrotechnics Symposia are available for purchase from the host organization, IIT Research Institute.
For prices and ordering information, contact:

Dr. Allen J. Tulis
IIT Research Institute
10 West 35th Street
Chicago, IL 60616

Information regarding availability, prices, and ordering of the Pyrotechnica publications can be obtained from:

Mr. Robert G. Cardwell
Editor and Publisher
2302 Tower Drive
Austin, TX 78703

APPENDIX B: MIXING TEST QUANTITIES OF PYROTECHNIC COMPOSITIONS

The pyrotechnic chemist always begins with a *very* small quantity of composition when carrying out initial experiments on a new formula. The preparation of one or two grams of a new mixture enables one to evaluate performance (color quality and intensity, smoke volume, etc.) without exposure to an unduly hazardous amount of material.

Eye protection -- safety glasses or goggles – is *mandatory* whenever any pyrotechnic composition is being prepared or tested. Necessary equipment includes a mortar and pestle, a laboratory balance, a soft bristle brush, several 2-3 inch lengths of fireworks-type safety fuse (available from many hobby stores), and a fireproof stone or composite slab on which to conduct burning tests.

Pre-grind the components *individually* to fine particle size. Do not grind any oxidizer and fuel together – fire or explosion may result. Weigh out the proper amount of each component and combine the materials in the mortar. Carefully mix them together with the soft brush to obtain a homogeneous blend. *Caution*: Do not prepare more than 2 grams of any composition for evaluation purposes using this procedure.

Place a small pile of the mixed composition on the fireproof board, insert a section of safety fuse into the base of the pile, and carefully light the end of the fuse with a match. Step back and observe the effect. Because of the generation of smoke by most pyrotechnic compositions, these tests are best conducted outdoors or in a well-ventilated area such as a laboratory fume hood. Be certain no flammable materials are near the test area, for sparks may be produced.

All testing of pyrotechnic compositions *must* be carried out under the direct supervision of a responsible adult well trained in standard laboratory safety procedures. Serious injury can result from working with larger amounts of composition or from the misuse of pyrotechnic mixtures, so caution and adequate supervision are mandatory. *Warning*: Do *not* attempt to prepare any of the explosive mixtures listed in Tables 8.5 or 8.6. These must be mixed only by remote means, or serious injuries might result. The color-producing compositions listed in Tables 7.9-7.13 are recommended as a good starting point for persons preparing their first pyrotechnic compositions. The effects caused by variations from the specified percentages can easily be seen upon burning.

INDEX